Challenges and Solutions in the Hydrogen Value Chain

Ankica Kovač
Editor

Challenges and Solutions in the Hydrogen Value Chain

State of the Art Perspectives on Decarbonization and the Green Energy Transition

Springer

Editor
Ankica Kovač
Faculty of Mechanical Engineering
and Naval Architecture
University of Zagreb
Zagreb, Croatia

ISBN 978-3-031-69050-1 ISBN 978-3-031-69051-8 (eBook)
https://doi.org/10.1007/978-3-031-69051-8

This Springer imprint is published by the registered company Springer Nature Switzerland AG
The registered company address is: Gewerbestrasse 11, 6330 Cham, Switzerland

Preface

To understand the *importance* of developing hydrogen technologies, it is best to clarify the consequences of not doing so, by emphasizing several influencing factors. The first is raising awareness about the danger of climate change. If we take 1850 (the beginning of the second industrial revolution) as a reference year, the global surface temperature has increased by more than 1 °C to date (some sources state as much as 1.38 °C), and the Paris Agreement aims to limit this increase to 1.5 °C. Another crucial factor is the advocacy of technologies that must be advanced in order to achieve the agreed on limit. *Climate change* is a problem for all of us globally and we all need to solve it, both collectively and individually now, regardless of how much someone contributed to carbon dioxide (CO_2) emission in the past.

- The production of *hydrogen* via *electrolysis of water* using *renewable energy sources* (RES) is a process without accompanying CO_2 emission, and its use in hydrogen fuel cells also ensures a process without CO_2 emission, thus realizing the circulation of hydrogen in nature without harmful emissions.
- The use of hydrogen as an energy carrier is inevitable in today's power industry based on RES, and very likely in future fusion power plants as well. By using hydrogen, there would be no greenhouse gas emissions and global warming would be slowed down. Along the way, tall chimneys would disappear, as well as all the chimneys of industrial power plants and neighborhood heating plants.
- It is also possible to have a significant impact on the design and construction of buildings, as well as pure cogeneration (simultaneous generation of electricity and thermal energy) and trigeneration (simultaneous generation of electricity, heat, and cooling energy) plants at the level of kilowatts, megawatts, and higher.
- Application in transport sector enables fuel cell electric vehicles with a much greater travel range, as well as the replacement of diesel locomotives on non-electrified lines.
- Application in the cement and artificial fertilizer industry also has high potential. Green/clean/renewable hydrogen plays a key role in balancing the power system based on RES.

- Hydrogen can be used as a medium for storing excess electricity in energy systems with a high share of intermittent RES. Therefore, the application of hydrogen is multiple, and it can be used wherever we need electrical, thermal, or cooling energy.

The key to a faster transition to RES is energy storage, specifically hydrogen storage. Hydrogen storage tanks of small (kg), medium (tens and hundreds of kg), and large (tons) capacity can be strategically geographically distributed and interconnected. They are the central part of the microgrid together with the systems of *RES power plant + electrolyzer stack + hydrogen storage tank + fuel cell stack.* By networking green microgrids, the entire geographic space of a region or country can be covered with one extremely flexible and robust green electrical network. It is also possible to directly connect the storage tank with a network of hydrogen pipelines. Hence it is possible to achieve that the entire energy and transportation system does not depend on fossil fuels.

Hydrogen is an excellent energy carrier and an excellent link to RES. Hydrogen as a fuel can be produced on-site anywhere where there is water and solar energy or wind energy or other RES, and stored in storage tanks for the days when we need energy. Hydrogen can be stored in them for an almost infinite time. Thus, the key is in energy storage. Whoever solves long-term (especially seasonal) energy storage will solve *energy independence.*

Having said that, this book brings *challenges* and *solutions* in the future energy systems based on hydrogen across *complete hydrogen value chain.*

As the Editor, I would like to express my sincere thanks to all the authors who contributed to the realization of this book. Thank you for your precious time invested in shaping and writing each chapter. The work with you was such inspiring and great pleasure.

And of course, my sincere thanks go to the entire Springer Editorial Team. Thank you for your continuous support, great patience, kind communication, and all your help from the very beginning of this journey.

Zagreb, Croatia Ankica Kovač

Contents

Considerations for a Scalable Business Model for Clean Hydrogen

Frank Wouters

1 Introduction

Renewable hydrogen will play a crucial role in a climate-neutral economy, as shown in many recent scenarios. In a system soon dominated by variable renewables such as solar and wind, hydrogen links electricity across time and space with industrial heat, materials such as steel and fertilizer, space heating, and transport fuels. At COP28, the United Nations climate conference in Dubai, representatives from nearly 200 countries pledged to transition away from fossil fuels, and renewable hydrogen combined with renewable electricity has the potential to entirely replace hydrocarbons.

However, despite a growing pipeline of hydrogen projects globally, as laid out in the International Energy agency (IEA)'s Global Hydrogen Review 2023 [1], **only 4% of this potential production has at least taken a final investment decision (FID).** Several factors contribute to this: a cost gap between current production costs and the cost of alternatives, a mismatch between offtake and supply in terms of contract duration, recent price increases due to interest rates and inflation, and complexity of support measures.

This chapter describes a hydrogen market introduction model that mimics the successful German Renewable Energy Act (2000) to bring hydrogen volume into the system while bridging the cost gap [2].

F. Wouters (✉)
MENA Hydrogen Alliance, Abu Dhabi, UAE
e-mail: frank@frank-wouters.com

A. Kovač (ed.), *Challenges and Solutions in the Hydrogen Value Chain*,
https://doi.org/10.1007/978-3-031-69051-8_1

2 The Renewable Energies Act

In 1999, a group of German parliamentarians led by green politician Hans-Josef Fell and socialist Hermann Scheer drafted the Renewable Energies Act or EEG (Erneuerbare-Energien-Gesetz). The EEG was both simple and revolutionary, perhaps because it was drafted by members of parliament and not professional law drafters from a ministry. Although initiated by the ruling red-green coalition, the EEG proved remarkably resilient even under the Christian Democratic rule of future governments. The EEG provided a mechanism through which anyone that constructed a solar PV project, wind farm, biogas digester, or other renewable electricity generator would receive a 20-year cost-reflective fixed feed-in tariff. This provided certainty for future cashflows, so banks could finance against that, once they felt comfortable with the technology. The local utility had an obligation to take the power and pay the tariff, which was higher than the average power prices at the time, but they could pass the cost differential on to the national high voltage grid. There, all the renewable power was pooled, and the financial burden equally spread over all German power customers, with some notable exemptions not to hurt power intensive industries and protect German competitiveness. The increasing share of green power in the power mix was made explicit in monthly electricity bills, so consumers knew what they were paying for. This system ensured fairness, because the wind-rich states at the coastal north saw a steep increase in wind projects, and this would have been a much heavier burden on them if not for this redistribution. Equally important, this was not a tax-and-spend system, as was the case in many other countries that tried to copy the German model. The European Court of Justice ruled in 2001 that the EEG was not anti-competitive, mainly because **it was not considered a subsidy. It was just a very elegant way to socialize the cost differential, with a decreasing gap over time, as renewable technologies became ever more competitive.**

When Angela Merkel took over the government in 2005, she quickly realized that the main initial beneficiaries were farmers, most of whom were her christian democratic union (CDU)'s constituents. Many in the north of the country installed wind turbines on their land, earning some extra cash without losing use of their land. In the south, farmers installed solar PV systems on their barn roofs. More importantly, though, the EEG was entirely budget-neutral but allowed Germany to become an early leader in the deployment of renewable energy, making it politically very robust. Of course, the German electricity consumers had to pay the bill somehow, and over time, opposition against the financial burden of the EEG grew. But many chose to install systems themselves, participate in community-owned wind farms, and personally benefit from the system.

It is fair to say that the German EEG unlocked the renewable energy revolution, with renewables now offering the lowest cost of power almost everywhere, also assisted by China's industrialization of solar PV production. Some other countries also introduced feed-in tariffs but failed to adjust the tariffs quickly enough to avoid windfall profits for developers when prices of PV panels and wind turbines dropped.

This led several countries, including Spain, Italy, and Czechia, to retroactively cut tariffs or introduce taxes, leading to lawsuits and a somewhat bad reputation for feed-in tariffs. Nonetheless, the simplicity of the scheme and its political robustness are beautiful examples of effective policies that have achieved the desired outcomes: cost-effective renewable electricity and increased energy security without a need for further subsidies. We now have more than 1TW each of installed solar PV and wind capacity globally. The EEG laid the foundation for that.

3 Electrons and Molecules

At the World Hydrogen Summit 2023 in Rotterdam in May 2023, Executive Vice President of the European Commission Frans Timmermans called hydrogen "inevitable.". I have argued before that the end game of the energy transition will be a situation where electricity's share of final energy demand will grow from the current 20% to 50%, and all of that must be green.

Electrification is the bedrock of the energy transition Not only are we greening and cleaning current electricity uses, but we are also using more and more electricity in areas traditionally dominated by fossil fuels, such as transport, industry, and space heating. The IEA's annual Global Electric Vehicle Outlook shows that more than 10 million electric cars were sold worldwide in 2022 and that sales are expected to grow in 2023 by another 35% to reach 14 million [3]. The year 2022 also saw a 40% increase in European deployment of heat pumps. According to IRENA, 28% of all electricity produced in 2021 was renewable, up from 21% in 2013 [4]. The IEA predicts that renewables and nuclear energy will dominate the growth of global electricity supply over the coming years, together meeting more than 90% of the additional demand [5].

Electricity is on a roll, and we have cracked the nut on generating it cleanly and cost-effectively. However, **electricity only constitutes slightly more than 20% of all final energy consumed globally today; the rest are molecules in the form of solids, liquids, or gases.** Most energy molecules we use are fossil fuels, i.e., coal, oil, and methane. We burn fossil fuels to propel vehicles, cook our food, make steam, and heat our houses, which causes CO_2 emissions leading to global warming and climate chaos. Even most electricity is currently produced from fossil fuels that are transported and stored as molecules at a power plant. The meteoric rise of renewables has made it a fundamental aspect of the energy transition, and some people dream of an all-electric future, where everything we do or operate is powered by electricity. This electricity would be endless and clean, generated by solar panels and wind turbines, stored in batteries and pumped hydro systems, and transported through wires, some even spanning the entire earth.

Let's delve deeper into specific facets of this dream.

3.1 Transmission

A recent US study found that a reliable power system that depends on high levels of renewable energy would be impossible to implement without doubling or tripling the size and scale of the US transmission system [6]. The wind blows hardest in the vast plains of the Midwest, yet the bulk of the power demand is on the east and west coasts. In most countries or regions, we will have to transmit large amounts of energy from where it is generated to load centers elsewhere. The challenge is that the existing grid infrastructure is currently inadequate to effectively transmit the increasing power output of the fast-growing renewable energy capacity to the load centers. Look at Germany as an example: In 2015, the cost of curtailing renewables due to grid congestion was approximately €402.5 million [7]. Fast forward to 2022, and this figure had surged tenfold to exceed €4 billion [8]. Despite ambitious targets set by the German federal government, the expansion of the grid hasn't kept pace with the rapid growth of renewables, leading to ongoing challenges. This issue can be attributed to a straightforward cause: a mismatch in pace of development. **It takes 1–3 years to develop and construct a wind farm or solar plant, whereas it currently takes on average 10.2 years to construct a high-voltage grid line in the US and Europe** [9]. People routinely resist the installation of overhead power lines, and the permitting procedure in many countries is often lengthy. On a planning level, transmission expansion is slow because it is difficult to predict exactly where future generations will be located, especially in liberalized and unbundled markets, who will benefit from it and who will pay for it. In a recently released report, the IEA identifies a large and growing queue of renewables projects waiting for the green light to be connected to the grid, pinpointing 1500 gigawatts worth of these projects that are in advanced stages of development. This is five times the amount of solar PV and wind capacity that was added worldwide in 2022 [10]. There are even indications that this number may be double that, i.e., 3000 GW, more than the entire global renewables capacity today. However, the speed of grid expansion, or lack thereof, is not the only problem.

There is a difference between electrons and molecules as energy carriers. The bulk transmission of energy is typically done via high-voltage cables for electrons and in pipelines for gases and liquids. It is worth comparing the two. The 1000 MW BritNed sea cable is a 260 km HVDC interconnector that links the UK electricity grid with the Netherlands. It was commissioned in 2011, and cost approximately €600 million [11]. The Balgzand Bacton Line, or BBL Pipeline, is a 36-inch natural gas interconnector, also between the UK and the Netherlands. The 235 km pipe was commissioned in 2006 at a cost of €500 million and an initial capacity of 16 billion cubic meter (bcm) of natural gas per year [12]. The BritNed electricity cable can carry 8.7 TWh in a year, whereas the 16 bcm natural gas pipe can carry the equivalent of 170 TWh. **So, for roughly the same investment, it is possible to transmit 20 times more energy in a pipeline for molecules compared to a cable for electrons.** This clear advantage of molecules over electrons is often overlooked. Now, the obvious elephant in the room is that natural gas is a fossil fuel and not a good comparison for an electricity cable that can transmit clean, renewable energy. Fortunately, most natural gas pipes can be modified to accommodate hydrogen, which can be made using renewable electricity (i.e., green hydrogen), at a fraction of the cost of newly built pipes. Since gas pipelines are already available in many places on the planet, and most of us are committed to decarbonizing our energy systems, we should explore converting this existing infrastructure to accommodate future renewable energy carriers such as green hydrogen. A group of European gas transmission system operators (TSOs) have started working on this, and they state they can use most of Europe's 200,000 km of high-pressure methane pipes for hydrogen [13].

3.2 Storage

Storage is another important element in any energy system. It is required to ensure stable energy supply to end users amid variations in energy production and demand. Molecules are easy and cheap to store—coal can be stored in giant heaps, oil and diesel in tanks, and natural gas in underground reservoirs, typically depleted gas fields, aquifers, or salt caverns. It is not possible to store electrons, so electricity must always be converted into some other form of energy and reconverted back to electricity when it is needed. Let's look at Europe to get a sense of the scale. The European Union's gas demand is currently about 400 billion cubic meter (bcm) per year, of which 100 bcm is stored because of the seasonal demand pattern and energy security considerations. This roughly equates to 1000 TWh. Europe's pumped hydro capacity, the only grid-scale electricity storage, stands at approximately 145 TWh. According to Ad van Wijk the investment costs of pumped hydro are more than 100 higher than for underground gas storage in an aquifer [14].

In a future energy system that is based on renewable energy, many places will show a seasonal imbalance between supply and demand patterns. More renewable energy can be generated in the summer, but the demand is higher in the winter due to increased demand for heating and lighting. Therefore, several thousand TWh of seasonal energy storage will be required, the bulk of which will have to be molecules due to the limited potential of additional pumped hydro storage, the limited feasibility of electrochemical storage at this scale, and the high costs associated with both.

It should be noted that even **the current electricity system consists of both transport and storage of molecules and electrons, with the bulk of the storage and transmission in the form of molecules until the power plant, after which electrons take over.**

Most authorities, including IRENA and the IEA, estimate the maximum share of direct electrification in a fully decarbonized system to be 50%. Beyond this level, it becomes increasingly difficult and expensive, especially regarding bulk long-range transport of energy and long duration energy storage. Molecules will then constitute the remaining 50%, of which at least half will be covered by green hydrogen. It must be noted that green hydrogen also requires green electricity to produce, so the overall green electricity input into this future system is at least 75%.

The alternatives to green hydrogen are biomass, CCS, or more electricity, and these options are inherently limited due to environmental issues, competition with food, limited geographic potential, and issues with infrastructure. The underlying assumption is that to avoid catastrophic climate chaos, at the latest by 2050, all energy-related emissions need to be zero.

3.3 Kick-Starting the Hydrogen Economy

Despite the promise of "inevitability" and necessity, the bulk of hydrogen announcements are currently still that: announcements. According to the IEA, annual production of low-emission hydrogen could reach 38 Mt. in 2030 if all announced projects are realized, although 17 Mt. come from projects at early stages of development [15]. The potential production by 2030 from announced projects to date is 50% larger than it was at the time of the release of the IEA's Global Hydrogen Review 2022. **Only 4% of this potential production has at least taken a final investment decision (FID)**, a doubling since 2022 in absolute terms (reaching nearly 2 Mt). Of the total, 27 Mt. are based on electrolysis and low-emission electricity, and 10 Mt. on fossil fuels with carbon capture, utilization and storage.

On the supply side, many of the current project proposals are in locations where land is cheap and available and the resource is good, which promises low production costs. On the demand side, there are various drivers for the use of clean hydrogen, some commercial, but many are driven by regulations such as carbon pricing, quotas, or obligations. On top of demand and supply, there are overarching targets laid down in national and regional hydrogen strategies, such as those laid down in the Hydrogen Accelerator of the European REPowerEU strategy, amounting to 20 million tons of clean hydrogen by 2030.

To bridge the current financial gap between green hydrogen and alternatives, many countries are introducing mechanisms, including India's production linked incentives, the USA's Inflation Reduction Act, and Europe's Hydrogen Contracts for Difference, and the recently established EU Hydrogen Bank, which could be used to Europeanize Germany's H2Global auction system.

What is lacking in all these schemes is simplicity and replicability In response to the USA's Inflation Reduction Act or IRA, which can add up to a subsidy of $3/kg of hydrogen as tax credits, the European Commission recently relaxed State Aid rules to allow member states to match funds a potential investor would receive in the US or elsewhere. To ensure that the aid will encourage a company to remain in Europe, cross-border investments must involve projects in at least three EU countries. The companies must also use state-of-the-art production technology to reduce environmental emissions. Such support, although generous on paper, is tailor-made and complex. In addition, this lottery-like system where governments select projects is inherently unfair, unlike the IRA or EEG, which was/is equally accessible to everyone.

In addition to financial support, the potential producer would also need someone who is willing to sign a bankable long-term offtake agreement enabling third-party financing of the investment. Producers and offtakers need to match volumes and timing, and all of this requires close coordination with the support scheme. Although there are some initial attempts to solve this conundrum, and Germany's H2Global is an interesting concept of double auctions and pooling of production and demand, there is inherent complexity in this approach.

Why don't we learn from the successful EEG?

4 A Feed-In Tariff for Hydrogen

As explained above, the current attempts to start a hydrogen market in Europe, the US, and other countries are all complex. **The beauty of the German EEG was its simplicity and replicability. Soon after introduction, the standardization of contracts, mortgages, and loan agreements further accelerated the market introduction. It was essentially scalable.** Let's think about a similar approach for hydrogen and propose a feed-in tariff for hydrogen.[1]

The basic principle is that any hydrogen producer can sell into a pool and does not need to worry about matching offtake volumes and pricing, and that the cost differential will be socialized over all energy consumers. On the offtake side, consumers will be offered clean hydrogen at a price level that will incentivize the switch away from current fossil energy carriers. A major difference between electricity and molecules, however, is that electricity feed-in tariffs have a national or regional grid to feed into. Hydrogen grids are currently not available, apart from a few limited exceptions. Hydrogen is currently mostly produced in captive installations, and traded volumes are limited. In the short term, such captive installations, mostly fossil fuel gasifiers or steam methane reformers, will be replaced by local electrolyzers. And on a regional scale, hydrogen valleys are being developed, while in parallel, several hydrogen backbones are being constructed. However, it will take a long time before a full hydrogen infrastructure is available that would allow for a similar physical feed-in mechanism used in the case of electricity. So, what is the solution to this infrastructure conundrum? **We will need to decouple the physical delivery of molecules from the virtual quality, which can be expressed in certificates.**

In the first phase, until the hydrogen production costs can match natural gas prices, a clean hydrogen production tariff mechanism is proposed with a hydrogen clearing pool to distribute the cost among all energy clients. Currently, clean hydrogen is more expensive than natural gas, even considering a carbon price.

[1]This was originally outlined by the author and Prof. Dr. Ad van Wijk in "Hydrogen Rocks – Towards the Creation of the European Hydrogen Economy" (ISBN-13 (15) 9789463663854).

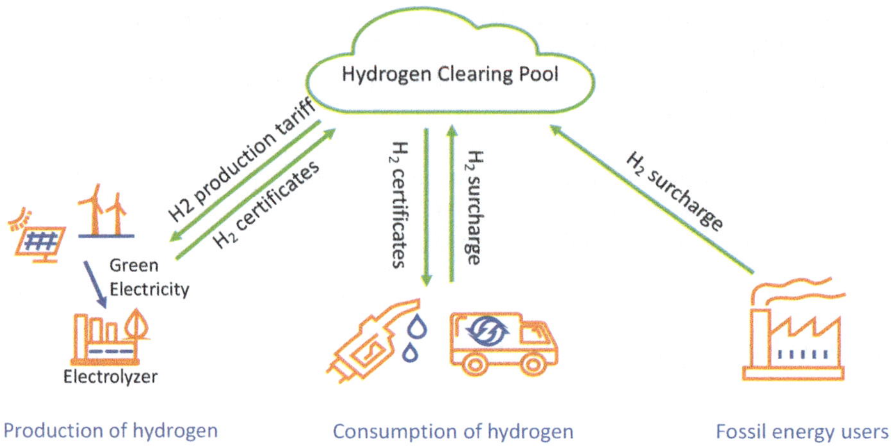

Fig. 1 Hydrogen clearing pool

However, most analyses predict renewable hydrogen to be the lowest-cost gas option by 2050 due to the learning curves of electrolysis and renewable electricity.

A Clean Hydrogen Production Tariff mechanism is proposed that can stimulate production and create hydrogen volume by pooling the initial cost gap for clean hydrogen in the period until cost parity for hydrogen in the market has been achieved. The cumulative payment for the cost gap is distributed evenly and fairly across all energy customers through a clearing mechanism.

Figure 1 shows the principle behind the hydrogen clearing pool mechanism. Producers of clean hydrogen will receive a Clean Hydrogen Production Tariff, which is fixed for 20 years and achieves a return of approximately 10%. The tariff is cost-reflective and differentiates between regions and system sizes. Larger systems receive lower tariffs due to economies of scale and lower production costs. Regions with lower renewable energy costs receive lower hydrogen production tariffs, allowing for equal regional spread of production across a country or region where the system applies.

Clean Hydrogen Certificates will be issued for units of hydrogen produced and will be exchanged at the national or regional hydrogen clearing pool in return for the hydrogen production tariff released by the pool. The pool is fed by a hydrogen surcharge on all energy consumed in the country or region, both clean hydrogen and fossil.

The concept is to socialize the costs through the above-described hydrogen clearing pool instead of using public funds, whether national or regional. That way, the difference between the cost of hydrogen and the sum of the CO_2 and fossil fuel prices, which is expected to diminish over time, is fiscally neutral and not dependent on national tax money, the availability of which is likely to be volatile and in competition with other worthy causes. The intent is to create a political robustness like the German EEG.

Certain natural gas or hydrogen-intensive sectors operating in an international, competitive environment may receive an exemption for paying the renewable hydrogen surcharge. Renewable hydrogen is expected to be competitive by 2035, and a clean hydrogen production tariff mechanism is then no longer necessary and could be replaced by auctions and tender procedures, as has been the case in Germany following the initial EEG. A review mechanism must be defined to evaluate the progress, provide a rolling forecast, and assess cost and price levels. A major review shall be carried out every 3 years, in line with current practices related to long-term natural gas contracts. However, a quarterly review is required to benchmark the tariff levels to avoid overpayment or underpayment.

A system of carbon contracts for difference (CCFD) as a possible policy instrument for scaling up hydrogen has been mentioned in the European hydrogen strategy. The idea is to link it to the Emissions Trading System and to pay a CCFD to a consumer to bridge the cost gap between green hydrogen and alternatives. A notable difference with that proposal is that I propose that the carbon contracts for the difference be paid to the hydrogen producer rather than the consumer. This crucially improves the bankability of hydrogen production projects.

There will be several distinctly different archetypical hydrogen ecosystems:

- Captive hydrogen systems.
- Standalone hydrogen applications.
- Hydrogen Valleys.
- Hydrogen backbones.

Each archetype will operate under different economic principles.

4.1 Archetype 1: Captive Hydrogen Systems

A typical captive hydrogen system is shown in Fig. 2 above. Examples of this are hydrogen production plants inside refineries, or the hydrogen production unit in an ammonia plant. They are currently mostly steam methane reformers (SMRs) that can easily be replaced by electrolyzers. The produced hydrogen is delivered directly to a continuous process, for example, the Haber Bosch unit in an ammonia loop.

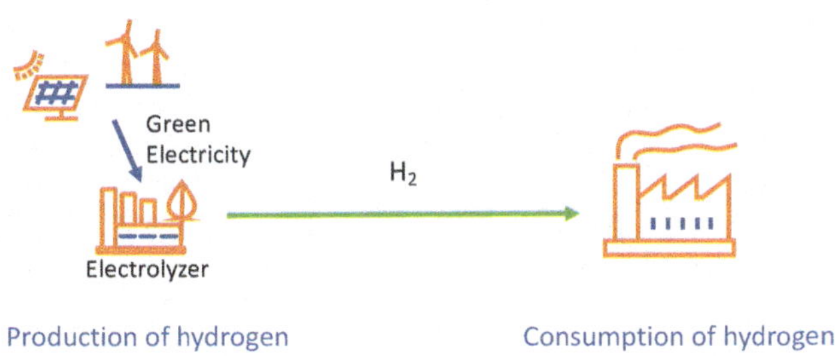

Fig. 2 Captive hydrogen systems

The easiest way to replace SMR units with electrolyzers is through obligations or quotas. By 2035, all SMR units should either have full CO_2 capture units attached, be replaced by electrolyzers fed by green electricity, or be connected to a hydrogen backbone.

4.2 Archetype 2: Standalone Applications

Figure 3 above shows a standalone hydrogen application. These applications differ from archetype 1 in that they don't replace conventional hydrogen but serve a new purpose, e.g., a hydrogen refueling stations with onsite hydrogen production.

The production of hydrogen is incentivized through the Clean Hydrogen Production Tariff system.

The price for the hydrogen charged in this archetype should be set to incentivize the switch from the currently used fuel to hydrogen. If hydrogen is used as a transport fuel, the price should be set to a level whereby the total cost of ownership expressed in €/km of the new transport modality is lower than the current alternative. If hydrogen replaces natural gas in a direct reduced iron furnace, that hydrogen should be cheaper than natural gas.

4.3 Archetype 3: Hydrogen Valleys

A hydrogen valley is a geographical area—a city, a region, an island, or an industrial cluster—where several hydrogen applications are combined into an integrated hydrogen ecosystem that consumes a significant amount of hydrogen. It should ideally cover the entire hydrogen value chain: production, storage, distribution, and final use. Figure 4 shows a typical set-up and components of a hydrogen valley.

Producers of hydrogen in a hydrogen valley sell their Clean Hydrogen Certificates to the national or regional hydrogen clearing pool and receive a 20-year fixed Clean Hydrogen Production Tariff. Consumers of hydrogen are treated in the same

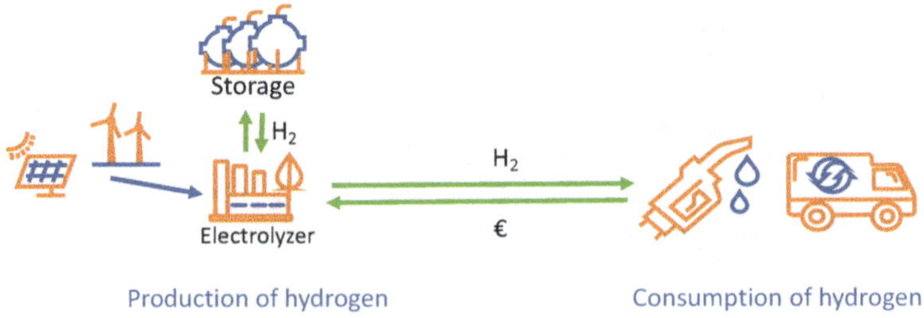

Fig. 3 Standalone hydrogen applications

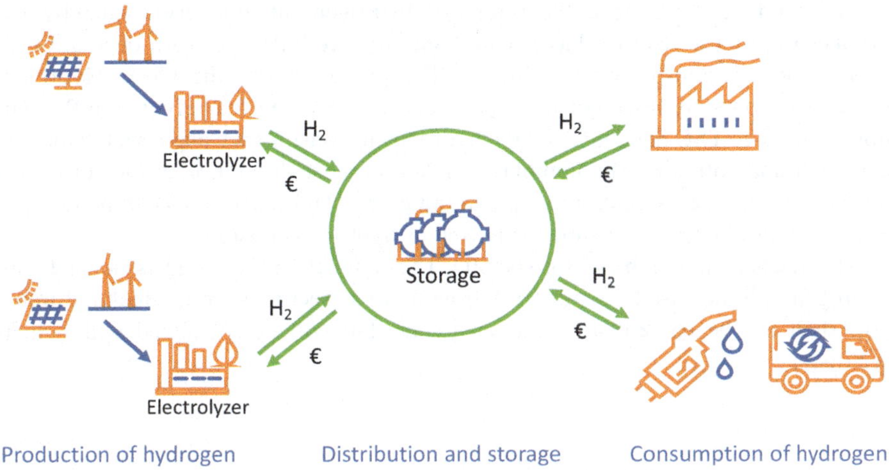

Fig. 4 Hydrogen valley

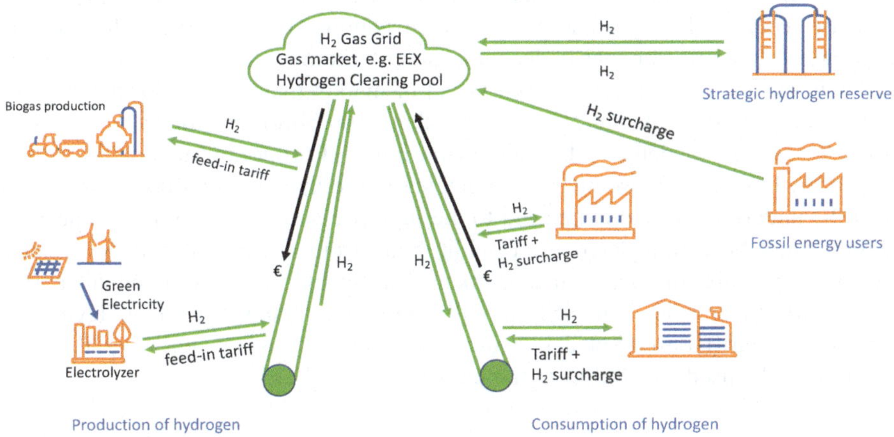

Fig. 5 A national or regional hydrogen grid and hydrogen clearing pool

way as in standalone applications, i.e., their hydrogen cost incentivizes a switch away from fossil alternatives.

4.4 Archetype 4: Hydrogen Backbones

Figure 5 below shows a national hydrogen grid as it could evolve in the medium term, with tariff mechanism, strategic reserve, and a hydrogen clearing pool.

The producers will feed the produced hydrogen into the grid, whereby the offtaker is either a TSO for large-scale high-pressure bulk hydrogen or a DSO for small-scale medium-pressure hydrogen. The producer sells the Clean Hydrogen Certificates to the hydrogen clearing pool and receives the Clean Hydrogen Production Tariff. The TSO and DSO sell the hydrogen to consumers using a sector specific approach and formula. The cumulative marginal cost difference in the form of a renewable hydrogen surcharge is spread out evenly and fairly across all natural gas and other fossil energy consumers, as well as hydrogen consumers.

The mechanism will be sector-specific. The cumulative surcharge associated with hydrogen volumes used by current non-natural-gas sectors such as steel and transport shall not be passed on to gas customers but will be socialized within their sectors.

5 A Strategic Hydrogen Reserve

The war in Ukraine has shown that strategic energy reserves are necessary. They increase security of supply and provide price stability. In the start-up phase of the hydrogen economy, they also decouple supply from demand, which will have different growth dynamics until some level of market maturity has been achieved. Let's look at Europe, which currently has about 100 billion m^3 methane storage—about 25% of annual demand. Using the same ratio, if Europe aims to have 20 million tons of hydrogen by 2030, then the strategic reserve must be 5 million tons, which shall be underground and connected to the high-pressure hydrogen backbone. The reserve shall be administered by a public body from a financial and governance perspective, while the actual management can be done by a commercial market player. Market players can sell into the reserve at the same Clean Hydrogen Production Tariff described above until the reserve is adequate. After that, the reserve can be used as a buffer in the market.

6 Imports

Hydrogen production potential as well as production costs vary significantly between regions. Many places in Africa and the Middle East, for example, are likely to export hydrogen and Europe, Japan, and Korea have announced they will need to import hydrogen in the future. Various sources have tried to estimate the amount of traded hydrogen across these regions, and the share for 2050 ranges from 20% [16], 33% [17] to more than 50% [18].

If we focus on Europe, with an ambitious hydrogen strategy and a declared intention to import hydrogen, the question is how to integrate imported green hydrogen into a European market with a central pool and socialization of costs as described above. It would be recommended to link such mechanisms as described in Article 6 of the Paris Agreement.

7 Book and Claim

The Clean Hydrogen Certificates will be issued, tracked, and redeemed in a book and claim model, which is common practice where a sustainability claim is separated from the physical flow of these goods. The most notable example is green electricity. Electricity cannot be tracked along the grid since it is all combined before entering a power outlet. To solve this problem, book and claim systems were developed to allow customers to claim a specific amount of renewable energy. Electricity providers can enter or "book" the electricity they have produced in their systems, and customers can "claim" the green energy they have bought. Consumers will then receive a certificate stating the amount of renewable electricity they paid for.

But what does this mean when talking about Clean Hydrogen Certificates? This means that hydrogen is not necessarily physically transported and entered into the specific backbone, storage or application that pays for green hydrogen. Instead, it goes into the nearest system close to the clean hydrogen production facility. The volume of clean hydrogen that is produced and entered into the system is tracked and verified, after which corresponding carbon emissions factors are calculated and accumulated at the hydrogen clearing pool level, where they are redeemed and turned into a surcharge to be spread-out over all energy customers on an annual level.

A national or regional registry will administer the clean hydrogen book and claim system and use a blockchain ledger to make the system tamper-proof.

Imports will be organized by semi-annual auctions as championed by Germany's H2Global. This allows for control of volumes, with the aim to have a maximum 50% share of imports.

References

1. IEA (2023) Global hydrogen review 2023. Paris, License: CC BY 4.0: https://www.iea.org/reports/global-hydrogen-review-2023. Accessed 31 Jan 2024
2. Federal Ministry for the Environment, N. C. (2000, March) Retrieved from Act on Granting Priority to Renewable Energy Sources (Renewable Energy Sources Act): https://web.archive.org/web/20160921012408/http:/www.bmub.bund.de/fileadmin/bmu-import/files/pdfs/allgemein/application/pdf/res-act.pdf. Accessed 31 Jan 2024
3. IEA (2023) Demand for electric cars is booming with sales expected to leap 35% this year after a record breaking-2022. Retrieved from https://www.iea.org/news/demand-for-electric-cars-is-booming-with-sales-expected-to-leap-35-this-year-after-a-record-breaking-2022. Accessed 31 Jan 2024
4. IRENA (2023) Renewable energy statistics. Retrieved from https://www.irena.org/Publications/2023/Jul/Renewable-energy-statistics-2023. Accessed 31 Jan 2024

5. IEA (2023) Electricity market report 2023. Paris, License: CC BY 4.0: https://www.iea.org/reports/electricity-market-report-2023/executive-summary. Accessed 31 Jan 2024
6. ESIG (2021) Transmission planning white paper. Retrieved from https://www.esig.energy/wp-content/uploads/2021/02/Transmission-Planning-White-Paper.pdf. Accessed 31 Jan 2024
7. Cleantechnica (2016) Re-dispatch cost Germany. Retrieved from https://www.cleanenergywire.org/factsheets/re-dispatch-costs-german-power-grid. Accessed 31 Jan 2024
8. Amprion. (2023). Amprion market report 2022–2023. Retrieved from https://www.amprion.net/Dokumente/Strommarkt/Marktbericht/2023/Amprion_Market-Report_2022-23.pdf. Accessed 31 Jan 2024
9. IEA (2023) Average lead times to build grid in EU and USA. Retrieved from https://www.iea.org/data-and-statistics/charts/average-lead-times-to-build-new-electricity-grid-assets-in-europe-and-the-united-states-2010-2021. Accessed 31 Jan 2024
10. IEA (2023) Electricity grids and secure energy transitions. Retrieved from https://www.iea.org/reports/electricity-grids-and-secure-energy-transitions. Accessed 31 Jan 2024
11. Reuters (2011) Britned. Retrieved from https://www.reuters.com/article/dutch-power-uk/update-1-britned-power-link-unites-uk-dutch-markets-idINLDE7301QJ20110401. Accessed 31 Jan 2024
12. GEMWiki (2022) BBL pipeline. Retrieved from https://www.gem.wiki/Balgzand%E2%80%93Bacton_Line_(BBL)_Gas_Pipeline. Accessed 31 Jan 2024
13. Marissa Moultak DT et al (2023) Assessing the benefits of a pan-European hydrogen transmission network. Retrieved from https://gasforclimate2050.eu/wp-content/uploads/2023/12/GfC_PanEU_230320_received_230323_published_final.pdf. Accessed 31 Jan 2024
14. van Wijk A, van der Roest L, Boere J (2023) Green energy for all—how hydrogen and electricity carry our future. Five Fountains B.V., Delft. isbn: 9789082763737. https://doi.org/10.5281/zenodo.8207489
15. IEA. (2023, September). Global hydrogen review 2023. Retrieved from https://www.iea.org/reports/global-hydrogen-review-2023
16. Deloitte (2023) Green hydrogen: energizing the path to net zero. Retrieved from https://www2.deloitte.com/content/dam/Deloitte/at/Documents/presse/at-deloitte-wasserstoffstudie-2023.pdf. Accessed 31 Jan 2024
17. IRENA (2023) Global hydrogen trade. Retrieved from https://www.irena.org/Energy-Transition/Technology/Hydrogen/Global-hydrogen-trade. Accessed 31 Jan 2024
18. Hydrogen Council (2023) Global hydrogen flows 2023. Retrieved from https://hydrogencouncil.com/wp-content/uploads/2023/11/Global-Hydrogen-Flows-2023-Update.pdf. Accessed 31 Jan 2024

Public Transit Zero Emissions Bus Applications: Battery and Electric Bus Comparison in a Canadian Case Study

Josipa G. Petrunić

Acronyms

AC	Alternate current
AES	Alectra Energy Solutions
AUC	Alectra Utilities Corporation
BEB	Battery electric bus
BEES	Battery electric energy storage
BRT	Bus rapid transit
BMS	Battery management system
CaaS	Charging as a service
CAIDI	Customer average interruption duration index
CEERP	Brampton's Community Energy and Emissions Reduction Plan
CI	Carbon intensity
CNG	Compressed natural gas
COS	Customer owned substation
CRU	Central rectification unit
CSO	Charging system operator office
CSV	Comma-separated values
CUSPAP	Canadian Uniform Standards of Professional Appraisal Practice
CUTA	Canadian Urban Transit Association
CUTRIC	Canadian Urban Transit Research and Innovation Consortium
DAC	Disadvantaged community
DC	Direct current
DEM	Digital elevation model
DER	Distributed energy resources

J. G. Petrunić (✉)
Canadian Urban Transit Research & Innovation Consortium (CUTRIC), Toronto, ON, Canada
e-mail: josipa.petrunic@cutric-crituc.org

A. Kovač (ed.), *Challenges and Solutions in the Hydrogen Value Chain*,
https://doi.org/10.1007/978-3-031-69051-8_2

DFAH	Self-contained clean-diesel-fired auxiliary heater
EaaS	Energy as a service
EIC	Energy infrastructure contract
EMS	Energy management system
EV	Electric vehicle
FCEB	Fuel cell electric bus
FDV	Future dollar value
GA	Global adjustment
GBE	Government business enterprise
GHG	Greenhouse gas
GIS	Geographical information system
GPS	Global positioning system
GTFS	General transit feed specification
GLT	Gas-to-liquid
GWP	Global warming potential
H_2	Hydrogen
HEB	Diesel hybrid-electric bus
HFS	Hydrogen fueling station
HOEP	Hourly Ontario energy price
HVAC	Heating ventilation and air conditioning
ICE	Internal combustion engine
ICIP	Invest in Canada Infrastructure Program
IESO	Independent Electricity System Operator
JIVE	Joint Initiative for Hydrogen Vehicles
kg	Kilogram
KPI	Key performance indicator
kW	Kilowatt
kWh	Kilowatt-hour
LCA	Life cycle assessment
LCEF	Low Carbon Economy Fund
LFP	Lithium iron phosphate
LH_2	Liquid hydrogen
LOTO	Lockout/tagout
LPG	Liquefied petroleum gas
MaaS	Microgrid as a service
MJ	Megajoules
MSA	Managed service agreement
MSP	Managed service provider
MSRP	Manufacturer's suggested retail price
MTO	Ministry of Transportation
NCA	Lithium nickel cobalt aluminum oxide
NMC	Nickel manganese cobalt
NMHC	Non-methane hydrocarbon
NPV	Net present value

NR	Non-revenue
NRCan	Natural Resources Canada
NREL	National Renewable Energy Laboratory
O&M	Operations and maintenance
OBC	Ontario building code
OCTA	Orange County Transportation Authority
OPG	Ontario Power Generation
OWSJ	Open web steel joists
P2P	Peer to peer
PM	Project management
PPA	Power purchase agreement
PPE	Personal protective equipment
PPRC	Peel Poverty Reduction Committee
RDD&I	Research, development, demonstration and integration
RFP	Request for proposals
RNG	Renewable Natural Gas
SAIDI	System average interruption duration index
SAIFI	System average interruption frequency index
SCM	Supply chain management
SLA	Service level agreement
SMR	Steam methane reforming
SOC	State of charge
TCO	Total cost of ownership
TTC	Toronto Transit Commission
V2G	Vehicle to grid
VKT	Vehicle-kilometres travelled
VOC	Voice of the customer
VOC	Volatile organic compound
VSM	Value stream mapping
ZEB	Zero emissions bus
ZeEUS	Zero Emissions Urban Bus System
ZETF	Zero Emissions Transit Fund

1 Introduction

On December 12, 2015, the United Nations Climate Change Conference (COP21) in Paris, France, declared the adoption of what is known today as *The Paris Agreement*, a legally binding international treaty on climate change [1]. This agreement was adopted by 196 parties, including Canada, and entered into force on November 4, 2016.

The overarching goal of *The Paris Agreement* is to hold "the increase in the global average temperature to well below 2 °C above pre-industrial levels" and "to limit the temperature increase to 1.5 °C above pre-industrial levels" [1]. In recent years, including at the COP28 Conference held in Dubai, United Arab Emirates, in December 2023, world leaders have underscored the need to limit global warming even more aggressively—to 1.5 °C by the end of the century. This updated goal is due to the raging consequences of climate change already evidenced worldwide by rising sea levels, increasing ocean acidity and uncontrolled, long-burning wildfires—exemplified by Canada's own abnormally extensive wildfire seasons from 2020 onward.

To limit global warming to 1.5 °C, greenhouse gas (GHG) emissions must peak by 2025 at the latest and decline by 43% by 2030 [1]. The result of these goals is the Government of Canada's suite of climate action oriented decarbonization funding initiatives, including the Zero Emissions Public Transit Fund (ZETF) discussed in this report. These targets and global efforts have resulted in Canadian cities declaring "climate emergencies" locally and initiating efforts to radically reduce their own operational emissions by 2030, 2040 and 2050, respectively.

Within this context in 2019, the Canadian Urban Transit Research and Innovation Consortium (CUTRIC) engaged in a series of predictive planning and implementation studies for zero emissions public transit buses.

The following study documents the predicted outcomes associated with an anonymous Canadian city, which owns a mid-sized public transit agency of approximately 400 buses. This city is exemplary and representative of Canadian public transit agencies that serve major cities—large enough to face the constraints of high-growth fostered by immigration to Canada but doing so within an environment in which sufficient public transit infrastructure has not been built out to serve the current needs, let alone ridership growth needs.

The City voted unanimously in 2019 to declare a climate emergency. The Council established an aggressive target to reduce the total GHG emissions generated in the city by 80% by 2050 [2]. The City's public transit agency plays a critical role in helping the City achieve this goal and has set itself an informal internal goal of reducing its total emissions to net zero, or as close to that as possible, by 2040 based on the results of the study conducted with CUTRIC. Throughout this report, 2041 is also referenced to account for vehicle rollouts in 2040. This timeline may be subject to change depending on the green transit market. Should supply chains for critical components such as zero emission buses or hydrogen fuel improve or lead to price point reductions in the immediate future, the timeline to full fleet decarbonization could be accelerated. However, if infrastructure construction delays, decreases in the availability of critical technologies, or other unforeseen circumstances, the deadline for full decarbonization may need to be extended. The Transit Agency will need to closely monitor the landscape for electrification and adjust its schedule accordingly.

The present study shows that achieving a net zero objective for the public transit agency by 2040 is possible though extremely challenging and costly, and it will necessarily involve hydrogen bus technologies as a core part of the decarbonization solution. Delaying the transition to 2050 may help to spread cost and risk factors over time, but it will not reduce the overall complexity of a full fleet transition.

This study identifies optimal mechanisms for decarbonizing the city's vehicle fleet while integrating new energy and fuel supplies. The analysis includes transitioning fossil fuel-based vehicles, such as diesel and diesel-electric hybrid buses, to fully zero emissions battery electric buses (BEBs) or hydrogen fuel cell electric buses (FCEBs), or a combination of the two technologies. The transition will require integrating new energy capacity and hydrogen fuel supplies into daily operations to energize the fleets.

To best assess the transition options available in the marketplace today, this report presents three scenarios for the electrification of the City's public transit fleet:

1. *Scenario One* assumes only BEBs are used to achieve decarbonization with in-depot and on-route charging ("full BEB solution").
2. *Scenario Two* assumes a transition using only FCEBs fueled by "green," "blue" or "gray" hydrogen ("full FCEB solution").
3. *Scenario Three* assumes a mixture of BEBs and FCEBs in the fleet design to minimize on-route charging episodes for BEBs while reducing the total fleet size ("mixed green fleet solution").

It is important to note that the scope of this study does not include a quantified assessment of "overall complexity." Today, the complexity of any decarbonized solution is determined agency-by-agency, case-by-case. Complexities vary by the specifics of each technology, such as how new or complicated a system or its fueling system may be, the complexity of the team labor involved and facility readiness. While a solution containing a mixed green fleet of different zero emissions bus types might reduce the number of buses required, it may also involve more complexity to manage two fueling types and their associated infrastructure. By comparison, a solution with only one type of decarbonized bus technology may reduce technological complexity but would require more buses and facility space compared to a mixed green fleet solution. In the case of FCEBs, maintaining the fleet may also involve managing a nascent and highly variable hydrogen marketplace in Canada, as the market is not yet ready to supply high volumes of hydrogen, leading to unknown complexities over the long term. Some complexities, such as provisioning electrical or hydrogen energy or maintaining vehicles and charging equipment might be offloaded to outside vendors to minimize complexity considerations through an Energy as a Service (EaaS) arrangement. However, these business models are also nascent and largely untested in the Canadian landscape.

Decarbonizing public transit is necessary but complicated. It will require an all-hands-on-deck approach. This new approach mandates a growth in new talent, new knowledge, new infrastructure, new routes and block designs and new thinking around how to design and deliver the best possible public transit service to riders today and into the future.

As one of the first agencies in Canada to establish a plan for full fleet decarbonization by mid-century, the public transit agency will experience unknown complexities in its mission. As a result, it will be a leader and a champion from which other agencies will learn over time. Today, the City must weigh the risks and benefits of each of the three electrification scenarios.

- Scenario One (full BEB solution) has the greatest emissions reduction potential.
- Scenario Two (full FCEB solution) has the highest total cost of ownership (TCO), but potentially less complexity, as it requires no charging infrastructure.
- Scenario Three (mixed green fleet solution) has the lowest predicted TCO and requires the smallest total fleet size but more complexity to integrate and manage two fuel types.

All three scenarios include a review of specific energy consumption outcomes that result from applying ZEBs to the public transit agency's specific blocks and depot arrangements. The fleet makeup assessed for all scenarios includes fleet growth up until 2040. The buses are either 12 meter (12 m), also called "40-foot buses," and 18 meter (18 m), also called "60-foot buses." This section outlines the varied outcomes of full fleet electrification across three scenarios.

1.1 Brampton Transit's Decarbonization Goals

The public transit agency under review in this study commenced operations in 1976 and currently operates 74 routes that connect to its neighboring municipalities. In 2010, the agency launched a Bus Rapid Transit (BRT) service, which provides limited-stop service along major service corridors throughout the city [5]. Before the COVID-19 pandemic, the public transit agency provided more than 31 million rides annually. Ridership decreased drastically during the pandemic but quickly recovered upon the reopening of the economy in 2022. The improved ridership resulted from lifting restrictions on mobility, socializing and workplace attendance by the federal and provincial governments. While some local service routes are still below pre-pandemic ridership levels, total system ridership is significantly higher than pre-pandemic levels.

As part of a fleet with depots at the two facilities the agency operates 476 buses, eight of which are BEBs [6]. The agency's fleet and average age are summarized in Fig. 1.

1.2 CUTRIC

The Canadian Urban Transit Research and Innovation Consortium (CUTRIC) is Canada's leader in technological and financial innovation and knowledge of decarbonized and electric transit buses. CUTRIC supports the commercialization of technologies through industry-led collaborative research, development, demonstration and integration (RDD&I) projects. CUTRIC brings innovative design to Canada's low-carbon smart mobility ecosystem, including zero and low emissions propulsion technologies and systems integration, smart vehicles and smart

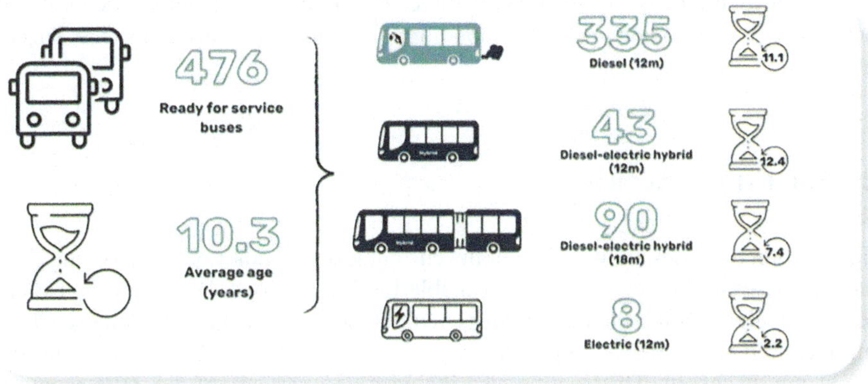

Fig. 1 Fleet and average age

infrastructure technologies, big data for mobility analytics and cybersecurity to advance mobility applications.

As a member-owned, non-profit technology consortium, CUTRIC has supported dozens of Canadian transit agencies in preparing their systems for electrification [7]. CUTRIC is proudly recognized as the Government of Canada's ZETF Fund National Planning Service, supporting public transit agencies to decarbonize their fleets [8].

1.3 Outline of the Analysis

The following section includes a detailed literature scan of the technologies assessed in this report, including battery-electric buses (BEBs) and hydrogen fuel cell electric buses (FCEBs). While other low-carbon fuel types are available in the marketplace today, such as renewable natural gas (RNG) and biodiesel, these fuels are not assessed in this study because they do not qualify for capital funding under the federal Zero Emission Public Transit Fund (ZETF).

Zero emissions buses (ZEBs) present several advantages over conventional diesel buses. They are often considered cleaner, quieter and potentially less costly to maintain over the long term than diesel equivalents. However, these technologies come with a suite of challenges. ZEBs are complex energy systems requiring high upfront investments in supporting charging and/or new fueling systems. Additionally, their actual maintenance costs over the long term are unknown. They require substantial labor force training, retraining and new skill sets. Section Two outlines these considerations.

The toolbox of methodologies used for this interdisciplinary analysis includes CUTRIC's non-profit (though proprietary) RoutΣ.i™ 3.0 simulation tool to help transit agencies improve fleet electrification planning and is used throughout this study, which includes an analysis of three duty cycles (i.e., light-, medium- and heavy-duty) to simulate different conditions of operation affecting the success of ZEBs.

2 Literature Review

This section provides an overview of the current global landscape for adopting ZEBs and discusses local incentives and available funding options. BEBs and FCEBs are reviewed, highlighting their advantages and potential application challenges. Charging technology for BEBs and hydrogen fuel generation and distribution considerations are discussed. This section also includes an overview of the ZEB economy, considering critical aspects such as vehicle maintenance and battery replacement costs.

2.1 ZEB Global Technology Landscape

Public transit improves the quality of life of Canadians by increasing road safety, reducing air pollution and benefiting large urban by supporting the economic mobility of their centres workers. While transit buses are a popular mode of travel in urban centers, primarily use diesel as a fuel source [10]. Burning diesel releases pollutants and toxic chemicals into the atmosphere and contributes to adverse health outcomes, including chronic diseases, respiratory illnesses and premature death [11–13]. Emissions from diesel engines also contribute to increased regional, national and global emissions of greenhouse gases (GHGs) [14]. An increase of only a few degrees in the global average temperature above pre-industrial levels can lead to extreme weather, water scarcity and exposure to vector-borne diseases, among other deleterious effects caused by extreme weather patterns linked to global climate change [15].

Public transit agencies transform the transportation and infrastructure industries toward zero emissions solutions. The widespread adoption of transportation electrification can mitigate the effects of climate change and adverse health outcomes associated with traffic-related air pollution.

Two technological solutions can significantly reduce emissions (to near zero) for public transit buses. The first solution uses batteries to store energy from the electrical grid, powering an electric motor for propulsion. This process is how a BEB electric vehicle (EV) operates. The second solution involves storing hydrogen onboard the vehicle and using the chemical energy in the hydrogen molecules to charge a small battery using a conversion device called a "hydrogen fuel cell." The fuel cell converts hydrogen into electricity. That electricity feeds an electric motor that propels a hydrogen fuel cell electric bus (FCEB).

An FCEB can be categorized as either "battery dominant," meaning the vehicle is propelled by electrical energy stored in the battery pack instead of electricity provided directly by the fuel cell stack, or "fuel cell dominant," meaning the vehicle is propelled by electricity directly emanating from the fuel cell stack without the need for a battery as an intermediary storage device, or with a very small battery serving as an intermittent storage device. The manufacturer's design solution depends on factors such as the relative sizing of the hydrogen storage tank compared to onboard battery capacity, the availability of space on the vehicle platform, the overall system's efficiency and the overall vehicle's cost and manufacturability. Hydrogen fuel cell buses sold in Canada today (of which there are very few) are primarily "battery dominant" models [16, 17].

In summary, BEBs are powered by electricity stored in rechargeable onboard battery packs. These buses receive most or all of their energy supply from a local electrical grid plugged for charging. Energy is stored onboard the vehicle in a battery pack. Energy from the battery pack powers an onboard electric motor that propels the vehicle [18]. Secondly, hydrogen FCEBs operate with a hydrogen fuel cell (HFC), which converts hydrogen stored in a tank into electricity. The electricity generated onboard charges a battery that powers an electric motor, which propels the vehicle [18].

2.1.1 Energy Production Techniques

The total emissions reduction potential for BEBs or FCEBs depends on the type of electrical grid used to power vehicles. In the case of FCEBs, emissions are determined by the type of hydrogen production technique used to produce hydrogen to fuel the vehicles.

Electricity can be generated through clean and renewable methods such as hydro, wind and solar or with low carbon intensity (CI) nuclear power. Replacing fossil-fuel vehicles with BEBs can be an effective solution for addressing emissions if the electrical grid generating the vehicle's electricity is a low-carbon grid system. As discussed in this report, the Ontario electrical grid is relatively green compared to other Canadian provinces. Ontario generates electricity through nuclear hydropower plants, natural gas, solar and wind. Therefore, ZEBs operating in Ontario would be relatively emissions free as Section Ten outlines.

In the case of FCEBs, the hydrogen stored onboard the vehicle's tank can come from two generation methods: electrolysis and steam methane reforming (SMR). The production of electrolytic ("green") hydrogen requires a significant amount of electrical energy. Emissions reduction would best use green hydrogen when the electricity production fueling its generation originates from a clean and renewable source. Green hydrogen has a low carbon intensity. Its source of emissions is the electricity generation on the grid. Hydrogen can also be made from steam methane reforming (SMR), which is significantly more carbon intensive. Given the low volume of electrolytic hydrogen available in Canada today, SMR-generate hydrogen (or gray hydrogen) is often much cheaper. Blue hydrogen is also produced from

SMR generation but includes a theoretical carbon mitigation strategy known as "carbon capture." Today, no blue hydrogen production supply chains in operation are available for public transit in Canada. Therefore, this type of hydrogen is referred to as "hypothetical" in this report, as it depends on whether carbon capture and storage (CCS) technologies can be demonstrated to work effectively to capture and store emissions over the long term—a feat not yet achieved in Canada [19].

Brief Global History of Decarbonized Technology Deployments in Public Transit

Both BEBs and FCEBs present advantages and disadvantages. Both technology types have been implemented globally with varying degrees of success. The most appropriate fleet solution for any public transit agency depends heavily on the operational characteristics of the transit system. Considering the marketplace supply of the fuel as well as the local energy matrix is crucial for determining the most cost-effective and environmentally beneficial approach. The following section provides a global literature review of these technologies worldwide and in Canada.

Electric buses are not new. Germany's electric trolley buses were in service as far back as 1882 [20]. The first trolley bus to operate in Canada entered service in 1921 in Windsor [21]. The dissemination of trolley buses was impeded in part because these vehicles receive energy through overhead catenary lines, which imposed a significant limitation on where they could operate. By contrast, contemporary lithium-ion batteries eliminate the need for catenary lines. In parallel, hydrogen fuel cell technologies have improved substantially over the years in their efficiency when converting chemical energy stored in hydrogen molecules into electrical energy. Because of these developments, electric buses can rely on an onboard battery for propulsion energy, or the combined energy from hydrogen electrolysis onboard. This, in turn, powers the battery onboard and propels the FCEB. Both technologies eliminate the need for catenary systems.

Transit agencies worldwide focus on fleet electrification using BEBs and FCEBs to garner environmental and economic benefits. In 2018, China claimed to be the biggest adopter of electric buses, reporting approximately 99% of all-electric buses worldwide, or about 425,000 vehicles [22]. China reports that Shenzhen is one of the few cities globally that has achieved full fleet electrification of its bus fleet, deploying more than 16,000 electric vehicles (EVs) [22]. In 2022, China's sales of new energy buses and trucks totaled 238,000 units, an increase of 90% from 2021 numbers. Out of these total sales, electric buses account for 58% (138,000 units), 97% of which are BEBs [23]. While these electric buses have avoided burning about 2000 barrels of diesel daily, GHG emissions of the country are not deeply affected due to China's reliance on coal to generate its electricity [24–26]. There are no means of independently verifying the accuracy of these statistics.

European efforts toward increasing the number of electric buses are notable given two large initiatives: the Zero Emissions Urban Bus System (ZeEUS) and the Joint Initiative for Hydrogen Vehicles (JIVE) [27, 28]. These initiatives include the contributions of pilot and demonstration projects, as well as commercial deployments from several European countries, including Austria, Belgium, Bulgaria, the Czech Republic, Denmark, Finland, France, Germany, Hungary, Israel, Italy, the Netherlands, Norway, Poland, Portugal, Romania, Serbia, Slovakia, Spain, Sweden, Switzerland and the United Kingdom. In 2021, 1.3% of the European Union's total bus fleet of 714,008 units was comprised of BEBs (approximately 9282 buses). The European BEB market continues to grow with BEB registrations of 4152 and 2567 units in 2022 and the first half of 2023, respectively [29, 30].

More than 3200 electric buses are reported to be operating in Latin America [31]. By 2025, 5000 new electric buses could be in service annually as cities across Latin America initiate their electrification pathway [32]. Chile currently leads regional efforts, reporting a fleet of at least 455 electric buses in operation and having announced a first electric bus factory with another pending procurement of 1000 additional electric buses [33–35].

North American cities recently began a fast transit electrification process, with the operation of about 650 electric buses on American roads in 2019 [36]. More than 1000 electric buses are running in the U.S., with almost 2000 more on order [37]. The demand for electric buses outpaces U.S. production capacity, indicating a strong market interest and commitment to sustainable solutions [36]. As of September 2022, the U.S. had 5480 electric buses in operation [38].

Globally, FCEBs are not yet as widely adopted as BEBs because of their higher upfront costs. In Europe, the JIVE project has deployed approximately 140 FCEBs and the JIVE 2 project plans to deploy another 152 FCEBs in 14 European cities [28]. The H2BusEurope project plans to place 600 FCEBs on the road [28]. In China, 1418 FCEBs were sold in 2018 alone of which 74 were purchased in preparation for the 2022 Winter Olympics [39]. Before the 2020 Summer Olympic and Paralympic Games, Toyota introduced 100 FCEBs in Japan, mainly in Tokyo, but no expansions were forecasted because of high capital and operational costs. South Korea's government envisioned deploying 1000 FCEBs by 2022, although the target was missed and is now expected to be reached by 2030 [40–44]. In the U.S., the National Renewable Energy Laboratory (NREL) in California currently monitors 64 FCEBs in operation with another 48 FCEBs in the planning stage [45]. Given that the State of California aims to achieve 100% of its transit fleet electrified by 2040, more FCEBs are expected to be deployed there in the immediate short term [46]. As of September 2022, the U.S. had 211 FCEBs in operation [38].

Concerns about whether batteries can deliver acceptable performance under extremely cold weather hindered early Canadian adoption of electric buses. After batteries were proven to perform adequately in a few demonstration projects, municipalities started to devise short- to long-term plans to ramp up electrification efforts [47, 48]. One example is TransLink in Vancouver, British Columbia, which currently operates four BEBs and two on-route chargers in the CUTRIC-led Pan-Canadian Electric Bus Demonstration and Integration Trial [49]. TransLink is acquiring 15 more BEBs to increase the fleet fivefold [49]. York Region Transit is another partner in this project and has deployed 12 BEBs. Brampton Transit also participates in the trial, operates eight BEBs and four high-powered chargers and has procured an additional 10 BEBs in late 2023 for early 2025 service [37, 49, 50].

Canada's decarbonization of public transit has developed significantly over the past 3 years. Several cities, from Vancouver, British Columbia (TransLink) to Halifax, Nova Scotia, have announced or recently procured new BEBs. Over the past 3 years, most announcements have been tied to the federal Zero Emissions Transit Fund (ZETF) which offsets the costs of zero emissions buses in Canada by up to 50%. Some recent highlights showcasing zero emissions bus (ZEB) procurements include the City of Ottawa committing to transit electrification by deploying four BEBs initially and recently approving the purchase of an additional 350 and two 150 kW pantograph chargers and four plug-in 150 kW chargers in depot [51, 52]. The Toronto Transit Commission (TTC) currently operates 60 BEBs procured from three different manufacturers—Build Your Dreams (BYD), New Flyer and Proterra [53, 54]. Several other cities and associated transit agencies have demonstrated similar levels of engagement, including the Toronto Transit Commission, Winnipeg Transit, Edmonton Transit Service, Société de transport de Montréal, Guelph Transit, Kingston Transit, London Transit and Greater Sudbury Transit, to name a few. These and other transit agencies have moved towards fleet electrification by either conducting pilot projects, purchasing electric buses or announcing electrification planning studies [55–61].

The FCEB landscape is much thinner. During the 2010 Winter Olympic Games in Canada, B.C. Transit operated 20 FCEBs in a demonstration project in the Resort Municipality of Whistler. At the time, it was the world's largest FCEB fleet demonstration at a single location. Despite the maintenance costs for these vehicles being reported as 58% higher than the maintenance costs associated with diesel vehicles, the project successfully diverted 4400 tons of carbon dioxide (CO_2) across 2.1 million km and 49 months of service [62]. In addition to intending to introduce 10 FCEBs, MiWay (Mississauga Transit) and CUTRIC are leading the Pan-Canadian Hydrogen Fuel Cell Electric Bus Demonstration and Integration Trial [63, 64]. Other cities across Canada are also considering using FCEBs for transit. Alberta has two FCEBs in a pilot program running in Edmonton and Strathcona [65]. Halifax Transit is conducting a feasibility study on incorporating 40 to 60 FCEBs into its service [66].

BEBs are Canada's most prevalent type of zero emissions transit vehicle, with the number increasing significantly in the past year [67]. Data collected through CUTRIC's ZEB Database™ also show that the distribution of ZEBs is not uniform across Canada, with some provinces and territories, such as Ontario and Quebec, leading the transition process. Others, meanwhile, have not yet adopted any ZEBs. This finding may suggest more incentives and funding programs to encourage greater participation of those agencies and provinces. The number of ZEBs in Canada is predicted to achieve the Canadian Government's Goal of introducing 5000 ZEBs by 2026.

2.2 ZEB Technical Overview

The two technologies under investigation in this study are BEBs and FCEBs. This section provides a deeper review of both technologies, followed by a literature scan of the technologies in global operations.

A BEB is a vehicle propelled by an electric motor that draws energy from an onboard battery. Efficient electrical motors providing the necessary power level for moving a transit bus have been feasible for the last century. Historically, the vehicles' energy storage capability and onboard energy systems have been key barriers impeding the widespread adoption of these vehicles [68]. The energy density of diesel, or the ratio of energy per unit of weight, is several hundred times greater than that observed in old battery technology. However, with the development of lithium-ion batteries, this multiplying factor has been reduced by approximately 100 [69].

Diesel vehicles still permit longer ranges in general, but BEBs are increasingly more feasible for various applications [70]. While some diesel buses can travel up to approximately 900 km without refueling, a BEB equipped with a large-capacity battery can travel approximately 300 km before needing charging in certain conditions. BEBs can also recover some of their kinetic and gravitational energy through regenerative braking, which helps to extend their range. Figure 2 shows example ranges for different bus types based on data from across Canada.

FCEBs are similar to BEBs because their propulsion is obtained from an electric motor that draws energy from an onboard battery. Compared to BEBs, however, FCEBs are equipped with a relatively small battery. In an FCEB, a fuel cell stack charges the batteries onboard. The fuel cell oxidizes hydrogen fuel to produce the electrical energy that charges the vehicle battery during operation. As a result of this operational dynamic, FCEBs can be considered BEBs equipped with "range

Fig. 2 National average ranges by bus type

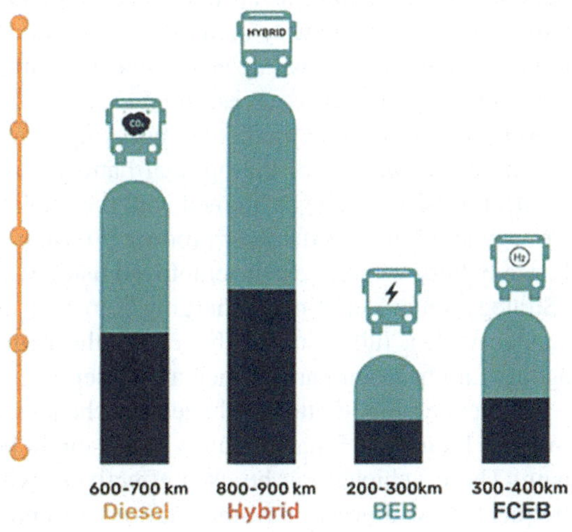

Table 1 National averages for bus comparison by fuel capacity, fuel efficiency and range[a]

	Fuel capacity	Fuel efficiency	Range
Diesel	400 L	60 L/100 km	600–700 km
HEB	400 L	45 L/100 km	800–900 km
BEB	500 kWh	1.7 kWh/km	200–300 km
FCEB	35 kg	10 kgH2/100 km	300–400 km

[a]Note: These values are approximate averages as specific circumstances result in variability

extenders." Since hydrogen fuel displays a much higher energy density than lithium-ion batteries, FCEBs can usually perform with significantly higher ranges than BEBs. FCEBs typically have a range of up to 400 km, which is higher than the range of BEBs but still significantly lower than the range of diesel buses.

Refueling the hydrogen tank of an FCEB is usually much faster than charging the battery of a BEB. Even though FCEBs possess a tailpipe, the exhaust gas is only water vapor, making the vehicle a ZEB [71]. Similar to BEBs, FCEBs can recover some of their kinetic and gravitational energy through regenerative braking.

Table 1 compares diesel, diesel hybrid, BEB and FCEB technologies. The ranges are subject to the number of passengers onboard, weather and road conditions. The ranges of BEBs and FCEBs are shorter than those of diesel or diesel-hybrid buses.

2.2.1 Review of BEB Vehicle Technology

BEBs with larger battery capacities can travel longer distances without recharging. However, procuring vehicles with a larger battery pack means incurring significantly higher capital investments, as the cost of batteries is a significant portion of the total purchase cost of a battery-propelled vehicle. A BEB with a larger battery is also a heavier vehicle that performs with decreased energy efficiency. The correct sizing of batteries is paramount, as an undersized battery can cause service failures. Meanwhile, an oversized battery increases costs and operational inefficiencies. Predictive modeling can help transit agencies select the most suitable battery size for specific operating requirements—an essential aspect of ZEB planning. Later sections of this report provide detailed predictive modeling outcomes to guide Brampton Transit.

The determination of the optimal battery type and capacity of a procured BEB (or fleet of BEBs) must be aligned with the public transit agency's intended charging strategy and bus manufacturer product offerings available in the marketplace today. Large-battery vehicles can be employed using what is described later in this report as "Strategy One: depot-only charging." In this charging strategy, buses start their service with a fully charged battery at the beginning of the day and only charge again at night when parked back at the depot.

However, small-battery vehicles are cheaper, lighter and less costly to refurbish over the long term. Smaller battery packs might be preferred for long-term economic gain. These vehicles can be used effectively when their service contains short but frequent downtimes at terminal stations to enable on-route charging episodes. In these cases, the vehicle still starts the service day with a fully charged (but smaller) battery, but it also experiences a few minutes of charging at a high-power charging system placed on-route several times during its daily operation. This case is called Charging Strategy Two (depot with on-route charging). Predictive modeling of an agency's specific blocks and routes can help select the optimal bus type and battery. Capacity and charging strategies for vehicles are required before significant investments in vehicle procurement or charging infrastructure are made.

One advantage of choosing Charging Strategy One (depot-only charging) is to avoid the capital costs of installing additional chargers to replenish the vehicle's battery on the route. These costs can be substantial because they can include the price of purchasing, installing and maintaining the equipment, as well as the eventual real estate costs for the locations where the equipment is installed.

Challenging blocks/routes might be difficult to electrify without high-powered on-route charging (also called "opportunity"). This difficulty is especially true for long blocks/routes, paths with high road grades, intense ridership and extreme temperature conditions, even when employing buses with large batteries. The introduction of on-route charging not only permits the operation of lighter and more efficient vehicles but can reduce the costs associated with power demand on the operator's electricity bill, depending on other factors, as analyzed later in this report.

Furthermore, vehicles with smaller batteries usually have a higher passenger capacity because less physical space is required for accommodating large battery packs. As a result, these vehicles can achieve higher operational efficiency levels [72].

Empirical evidence, mostly under idealized testing conditions (i.e., laboratory settings), shows that typical 12 m BEBs require between 1 and 1.5 kWh/km of energy for each kilometer of operation [73–75]. In real operational conditions, however, energy consumption can often surpass the 2 kWh/km mark depending on factors such as passenger load, use of auxiliary systems to power heating, ventilation, and air conditioning (HVAC) systems, starting and stopping frequency, and the route's topography. For these reasons, detailed planning is essential. Predictive modeling can significantly increase the chances of success, given that each portion of the fleet is assessed according to its difficulty level for electrification.

The advantages of fleet electrification go beyond the avoidance of harmful gas emissions. Electric buses are expected to require less maintenance than diesel or compressed natural gas vehicles because electric power trains employ fewer moving parts than internal combustion engines. In summary, electric buses are more energy-efficient compared to diesel fueled vehicles for several reasons [76]:

1. The peak efficiency values for internal combustion engines are approximately 45%, whereas electric motors can continuously operate beyond the 90% level and reach over 97% efficiency peaks [77, 78].
2. Electric buses have a regenerative braking mechanism that improves the vehicle's energy efficiency. Conventional fossil-fuel vehicles waste their kinetic and gravitational energy by dissipating it mainly as heat generated by the friction of the

conventional braking process. Electric buses can recover a fraction of that energy by allowing the electric motor to act as a generator, recharging the battery when the vehicle is slowing down or going downhill [79].
3. The analysis of empirical data gathered from vehicle testing, including electric, diesel, compressed natural gas (CNG) and HEBs, shows that electric buses are three to four times more energy-efficient than other vehicles overall [80].

2.2.2 BEB Charging Technologies

As discussed later in this report, Scenario One (full BEB solution) and Scenario Three (mixed green fleet solution) require ongoing, consistent and reliable charging solutions to succeed.

Three types of charging technology are currently available for BEBs: low-powered plug-in charging, high-powered overhead charging and mid-powered inductive charging.

1. *Low-powered plug-in* chargers can work in compliance with the Society of Automotive Engineers (SAE) EV and plug-in hybrid EV conductive charge coupler, commonly known as the J1772 type charger or "J Plug." The J1772 standard permits single-phase alternate current (AC) charging up to about 19 kW and direct current (DC) charging up to 400 kW. The chargers can comply with the J3068 standard, which uses three-phase AC electrical power instead and permits up to 166 kW of charging power. These two standards are interoperable. A BEB with a 600-kWh battery might take more than 4 hours to fully charge if initially plugged in with a significantly depleted battery. The longer charging time makes the charger more suitable for overnight charging at the depot [81–83].
2. *High-powered overhead chargers* usually conform to one of three sub-standards of the SAE's J3105—J3105-1, J3105-2 or J3105-3, depending on the connection configuration. J3105-1 requires an infrastructure-mounted cross-rail connection, J3105-2 requires a vehicle-mounted pantograph connection, and J3105-3 requires an enclosed pin-and-socket connection. J3105-1 and J3105-2 can reach up to 600 kW with Level 1 and up to 1.2 MW for Level 2, while J3105-3 is suitable for power transfers rated from 32 kW to above 1 MW. These high-power ratings allow a vehicle to obtain enough energy to complete a round-trip between terminal stations in roughly 5 min. High-powered overhead chargers are best suited for on-route charging. The chargers can be used in the depot when the vehicles do not have enough downtime for slow charging, or the charging schedule does not allow longer charging times [83–85].
3. *Mid-powered inductive chargers* allow wireless charging and do not require direct contact between chargers and buses. The charging power level can reach 450 kW, significantly reducing the charging time compared to plug-in chargers [86]. Inductive chargers can be used at a depot for overnight charging or at bus stops for on-route charging [87]. The equipment does not require much physical space, imposing no visual obtrusions [87, 88]. The use of inductive chargers is not widespread but may become more popular as technology advances.

Many variables, ranging from charging schedules to available physical space, determine whether to choose a single or a combination of charging technologies. Equipment compliant with the above standards should be safe and interoperable.

A primary challenge associated with implementing BEBs is their long charging times. To address this impediment, public transit agencies can deploy chargers capable of delivering energy at a high-power level at the depot or on-route to enable vehicle service delivery efficiently.

Chargers delivering large power levels can be leveraged to permit shorter charging times, but vehicles cannot accept an arbitrarily large amount of power. If a BEB is rated for charging at a given power level, it will not consistently be able to charge at that rate. The physicochemical properties of lithium-ion batteries are such that when the state of charge (SOC) is low, the battery will accept energy flowing at a more significant rate. When the SOC is high (greater than 80%), the battery's electrical impedance will rise, and the charging power level will decrease despite the power rating of the charger itself [89].

In addition to the internal physical properties of batteries, an onboard battery management system (BMS) will often limit the current flowing into the battery, especially if the system temperature increases and crosses a certain threshold established by the manufacturer. This power level management is often necessary to ensure the equipment endures an acceptable, useful life. Transit agencies should expect longer charging times on extremely hot days, especially when high-power levels are employed [90].

2.2.3 Challenges Impeding BEB Deployments

One of the primary challenges associated with adopting BEBs is the high capital costs associated with their implementation. A 12 m BEB with a battery capacity of 500+ kWh can cost approximately $1.6 million (2023 dollars), approaching double the price of a diesel-powered vehicle (estimated at approximately $900,000 in 2023 dollars). Due to the range limitations of BEBs, for most large transit agencies, it would likely not be possible to electrify a whole fleet on a one-to-one vehicle replacement basis. This reduced range results in even higher upfront costs associated with full fleet electrification relying solely on BEB technologies because of the likely need to procure 10% to over 50% more BEBs compared to diesel buses. Additionally, procuring and installing the required charging infrastructure to support these vehicles can further elevate upfront decarbonization costs for public transit agencies. One low-powered plug-in charger can cost anywhere between $50,000 and $500,000, and one high-powered overhead charger ("on-route charger") may cost approximately between $750,000 and $1.6 million, depending on civil works required [91]. On-route chargers typically incur additional installation, support, maintenance, and land acquisition costs. These increased capital costs pose a significant challenge for transit agencies pursuing transit electrification with BEBs. A later section in this report reviews the specific replacement ratios and infrastructure costs associated with a full and/or partial BEB solution, i.e., Scenario One (full

BEB solution), Scenario Two (full FCEB solution) or Scenario Three (green mixed fleet solution).

Despite the considerably higher upfront capital investments, BEBs can still present a lower total cost of ownership. Their operation is expected to be cheaper over the long-term because of savings on fuel and maintenance [45, 92]. Additionally, capital costs associated with electrification can be partially offset as certain manufacturers have or may introduce leasing options. For example, Proterra (a now bankrupt manufacturer) introduced a battery leasing program in its early years of operation. Another foreign bus manufacturer, BYD, currently provides an electric bus leasing program [93, 94].

A final capital cost consideration is the presumed rapid reduction in lithium-ion battery prices. As batteries currently represent a significant portion of the cost of BEBs, more affordable batteries may make BEBs increasingly competitive with diesel vehicles. However, this reduction may or may not arise in the short term, given global supply chain constraints exacerbated by pandemic conditions, which still affect battery supply chains in Canada. These supply chain constraints may even lead to price point increases in battery technologies in the short to medium term, leading to increased, rather than decreased, costs for decarbonizing fleets. If rapidly rising demand outpaces battery supplies and manufacturing capacities for heavy-duty applications, price-point increases may extend into the medium- and long-term future across the global marketplace that Canadian bus manufacturers and suppliers operate within.

Upfront costs that are higher compared to comparative fossil fuel equivalents, such as diesel, diesel hybrid-electric buses (HEB), and compressed natural gas (CNG) buses, are not the sole factor preventing broader and faster adoption of BEBs. Another challenge is the limited range of these vehicles. Even with a large 500+ kWh battery, BEBs may only achieve about 300 km in range or less if and when operating in harsh conditions or relying solely on charging at the depot. Diesel buses have a considerably longer typical range of 600–700 km, while HEBs have ranges of 800–900 km. Diesel buses are not as sensitive to variations in operational conditions, such as operating along a difficult road topography or with higher passenger and auxiliary loads.

Even though using on-route charging can significantly improve the range of BEBs, transit agencies will most likely be required to redesign schedules and blocks to include on-route charging time. To maintain existing service frequencies and satisfy service demands, the transition to fleet electrification will require additional capital assets (buses), a change in how service is planned and scheduled, and increased operating costs. These increased operating costs include, but are not limited to, additional charging/fueling time and increased service and operator hours. Based on CUTRIC modeling, ZEB fleets may see an increase in labor hours between 2% and 11%.

Depot charging schedules are a final consideration in the electrification process. Deploying many large-battery BEBs in a system can make charging an entire fleet overnight logistically complex or even impossible. If charged at a lower power level, each bus can take several hours to replenish its battery, thus requiring several chargers or a few charging stations equipped with multiple dispensers to charge multiple vehicles simultaneously. Alternatively, operating fewer chargers with higher power levels can be a viable alternative, depending on other garage operations and the labor availability required to move vehicles from chargers to parking. Depot automations that move vehicles from charger-to-charger or charger-to-parking, for example, on conveyor belt technology, would also influence charging strategies. In addition to lowering power demand charges on a depot/facility's electricity bill, on-route charging can augment depot charging operations since the vehicles would return to the depot with their batteries less depleted on average throughout the day, thus requiring less overnight or in-depot charging time in general.

In summary, higher upfront capital costs, range limitations, charging schedules, depot designs and labor schedules constitute several of the core constraints and challenges facing transit agencies seeking to transition from diesel buses to decarbonized solutions.

2.2.4 Review of FCEB Vehicle Technology

Hydrogen fuel cell electric buses (FCEBs) are usually assumed to have a greater range than BEBs on a vehicle-to-vehicle comparative basis. In some cases, they are presumed to have ranges equivalent to diesel buses. However, this comparable range is not always true.

The range of an FCEB compared to BEB, for example, is highly dependent on a BEB charging strategy, which might extend BEB ranges well beyond those of FCEBs in daily operations by allowing for continual top-ups and energy transfer throughout the day. FCEBs can compete with on-route BEB charging strategies *if* they have access to refueling during the day, which may be required. In the instance of a diesel bus comparison, FCEB ranges tend to be lesser than those of diesel bus equivalents based on products available in the Canadian marketplace today (this is largely driven by the size of the hydrogen fuel tank), and they are highly dependent on operating conditions.

Onboard an FCEB, a battery pack powers an electric motor like a battery electric bus. As described earlier, FCEBs also carry an additional energy conversion mechanism—a hydrogen fuel cell stack that converts hydrogen stored in a tank into electricity onboard. This additional energy conversion process makes them less energy efficient compared to BEBs as overall energy systems. Within these platforms, energy loss occurs in generating hydrogen, and further losses arise when converting the chemical energy stored in hydrogen molecules into electricity. A typical hydrogen tank installed on an FCEB can carry just under 40 kilograms (kg) of hydrogen fuel. A typical FCEB in Canada will consume about 10 kg/100 km, producing a standard FCEB range of approximately 400 km. According to the National Renewable Energy Laboratory in the U.S., optimal operations of FCEBs can enable them to travel anywhere from 320 to 560 km, depending on the duty cycles used [95]. This range is higher than most BEBs currently demonstrate and closer to the range of diesel or CNG vehicles (600–700 km). FCEBs impose fewer operational constraints than BEBs but cannot outperform diesel or compressed natural gas (CNG) buses in range.

The hydrogen refueling process takes approximately 10 min allowing FCEBs to be refueled during their daily maintenance service [96]. A more extended range and much shorter refueling time make FCEBs appealing to the transit electrification landscape. A typical FCEB can carry 40–60 kg of hydrogen, and current refueling technology enables the transfer of up to 10 kg of fuel per minute [97].

Depending on the hydrogen fuel tank's capacity and the onboard battery's capacity, a FCEB may be classified as either "fuel cell dominant" or "battery dominant." FCEBs sold in Canada today are considered "battery dominant," as their propulsion is powered by battery packs onboard rather than directly from the fuel cell stack. "Fuel cell dominant" vehicles are more prevalent in Europe.

The size and distribution of the refueling infrastructure for FCEBs are similar to that of their diesel counterparts. Hydrogen bus technologies avoid specific problems created by installing on-route chargers for BEBs. However, hydrogen infrastructure requires more space compared to diesel counterparts. This has been an important issue for transit agencies seeking to decarbonize with FCEB technologies, especially those near large urban centers where physical space is generally a significant constraint. The main impact of installing a hydrogen fueling infrastructure would be the loss of a few parking spots in facilities with parking spaces used at capacity.

A typical daily BEB battery SOC profile displays a downward trend as the vehicle consumes energy during operation, except for brief on-route charging episodes. FCEBs, on the other hand, usually display a steady SOC, whereby the battery SOC fluctuates around its optimal operational level if hydrogen fuel is available in the tank. This fluctuation happens because the fuel cell feeds the battery at a power level that approximately matches the power required from the battery to operate the vehicle.

While it may be theoretically possible to use FCEBs for a limited time after the hydrogen tank is depleted by relying solely on the energy stored in the battery, FCEBs are not designed to allow the full depletion of the battery for such operations to prevent battery degradation damage, and it is operationally inefficient to bring the battery state-of-charge (SOC) back to its optimal level afterward.

2.2.5 Review of FCEB Refueling Technologies

Scenario Two (full FCEB solution) and Scenario Three (mixed green fleet solution) require an ongoing, consistent and reliable hydrogen fuel supply to be successful.

Hydrogen is a fuel with a high gravimetric energy density, presenting approximately 120 MJ/kg compared to 45 MJ/kg for diesel. This difference means that the added weight when refueling with hydrogen is relatively small for a given amount of onboard energy. Nevertheless, the volumetric energy density of hydrogen is relatively low unless stored as a gas under high pressure (350–700 bar) or as a liquid in cryogenic reservoirs (−253 °C) [98]. Extra energy is necessary to compress hydrogen and reduce its stored volume. If stored in gaseous form at the highest-pressure level (700 bar), the volumetric density is about 42 kg/m^3. If stored in the liquid physical state, its density is about 71 kg/m^3 [99]. As more molecules can be stored in the same space with liquid hydrogen, it tends to present a lower transportation cost, even though liquefying the fuel requires cryogenic operations that incur additional costs, passed on to the end consumer (i.e., the public transit agency customer).

Hydrogen atoms can be found in abundant compounds such as water (H_2O) or methane (CH_4). However, these atoms must be separated from the compound into its molecular form (H_2) for use in fuel cell electric vehicles [8]. This separation process generally requires the input of a significant amount of energy, which can come from various sources such as burning fossil fuels, nuclear energy, and renewables such as biomass, wind, solar, geothermal and hydroelectric [100]. As a result, the environmental impacts and energy efficiency associated with hydrogen depend on its production method overall.

The three common methods of hydrogen production are electrolysis, producing "green" hydrogen: steam methane reformation (SMR) with or without carbon capture and storage technologies (CCS), producing "blue" and "gray" hydrogen, respectively [89]. Likewise, fossil-based natural gas, the biogenic RNG, can be further processed into hydrogen fuel through SMR, as discussed in Sect. 2.2.5. This process would reduce, however, the overall energy efficiency and potentially result in more GHG emissions compared to burning the RNG in natural gas buses [101]. Depending on the method of production, this method could have very low to even negative CI.

Green Hydrogen Green hydrogen is obtained through electrolysis by breaking up water molecules induced by an electrical current under certain physical conditions, producing a harmless oxygen gas. If the electricity used in this process is generated with a low-carbon intensity (CI) factor, such as in Ontario's very clean electricity grid, the electrolytic production method will result in substantially fewer carbon emissions than SMR-generated hydrogen (blue or gray) discussed below. This is why electrolytic hydrogen is often called "green" hydrogen.

Blue Hydrogen Blue hydrogen is produced from steam methane reformation (SMR) with carbon capture and storage (CCS) or carbon mitigation processes implemented to capture the resulting carbon emissions and store them to reduce the overall carbon intensity of the fuel. The carbon emissions of SMR-produced hydrogen are generally higher than those of green hydrogen but can be reduced by capturing and storing a portion of the carbon by-product. This capture usually occurs by either compressing or cryogenically liquefying the carbon dioxide by-product for injection into geological formations or burying it [102]. Hydrogen produced with SMR, including carbon capture (SMR-CCS), is often called "blue" hydrogen [103]. Last year, the United Kingdom announced the world's first at-scale hydrogen production facility that utilizes carbon capture and storage (CCS) [102] [102]. It should be noted that blue hydrogen is not currently being produced in Canada and will not be available in Ontario for the foreseeable future.

Gray Hydrogen Gray hydrogen is produced from a process known as steam methane reformation (SMR) with no carbon capture or carbon mitigation process. SMR-produced hydrogen without CCS usually presents the highest CI levels and is called "gray" hydrogen [103]. Gray hydrogen is also known as "dirty" hydrogen and is primarily used at scale within heavy industry.

Green hydrogen typically carries the lowest CI, but it is usually the most expensive to produce, as electrolysis is an energy-intensive process. Transportation of green hydrogen fuel also represents a significant portion of fuel costs if pipelines are unavailable.

A present and scalable solution to support the public transit agency's immediate climate action goals includes obtaining hydrogen fuel from a continuous supply chain that has scaled enough to significant volumes to deliver a large quantity of hydrogen fuel daily.

SMR hydrogen can benefit more easily from pipeline and storage infrastructure already available in Ontario and across Canada for natural gas. Technology standards are still evolving to permit the fuel to be safely transported in tank trucks as a gas with higher pressure values or as a liquid to minimize freight costs [104].

Globally today, hydrogen can be delivered through pipelines or trucks with tube trailers. The pipeline delivery method is the most economical, but the capital investment to build the necessary infrastructure can be prohibitively high. Hydrogen fuel can be carried in gaseous or liquid form when transporting hydrogen through trailers. Transporting gaseous hydrogen through trailers is only economical for relatively small-scale fuel demand. The liquid form of hydrogen is more economical when delivering enough hydrogen fuel to power a fleet of several hundred buses [105].

Localized hydrogen production is possible, as well. Site assessments for electrolyzers installed on site are not part of the scope of this study. Still, it is reasonable to conclude that a midsized transit agency would require substantial infrastructure in the form of electrolyzers and decompression systems to produce and distribute enough fuel to support the operation of several hundred FCEBs, and this infrastructure would occupy a material amount of real estate. If the public transit agency were to operate a liquid hydrogen storage facility, the agency would have to install liquid hydrogen storage tanks, cryo-pumps, vaporizers and fuel dispensers. This infrastructure can cost up to $5 million for a capacity of 1000 kg/day, which may support the operation of up to 30 FCEBs [106].

2.2.6 Challenges Impeding FCEB Deployments

Despite the range of advantages associated with FCEBs, these vehicles are less widely deployed in North America and Europe than their BEB counterparts. At the time of writing this report, only two FCEBs operate on Canadian roads in the City of Edmonton on a pilot basis.

Key reasons impeding the uptake of FCEBs widely in Canada as primary decarbonization technologies are listed below.

1. The purchase price of 12 m FCEBs is significantly higher than that of BEBs or HEBs. FCEBs can cost more than $2.00 million per unit or approximately 20% more than BEBs.
2. Green hydrogen fuel is limited in supply in Canada today and is relatively expensive compared to diesel and electricity. In Ontario, small-volume hydrogen supplies can cost as much as six times the price of an average liter of diesel, and the volumes under production are highly limited with no demonstrated green hydrogen delivery and supply chain, despite provincial and federal investments in hydrogen hubs in recent years. The ramp-up of Atura Power's (Ontario Power Generation) hydrogen hub proposed for the Niagara Region may gradually alter this landscape in the future, but, at present, volumes remain low and distribution remains untested. Gray hydrogen does exist in scaled supply chains in Ontario, specifically in Sarnia, Ontario, but it has never been leveraged to supply public transit agencies in Ontario or beyond given the high prices (still widely priced at $18–$70/kg and significantly more than diesel fuel) and for no anticipated environmental LCA GHG reductions [107].
3. Operating FCEBs is less energy efficient than operating BEBs from an energy systems standpoint. While this is more of an academic consideration at this point, it becomes an operational issue for cost controls and total GHG reduction planning at scale. When electrolytic hydrogen is used, the electrolysis process in a reasonably large power plant might only reach an efficiency of about 70%, meaning 30% of energy is lost in the fuel production method. Converting that hydrogen back into electricity onboard an FCEB also results in conversion losses. A fuel cell onboard a bus might only harvest 70–80% of the hydrogen energy content for propulsion. This means the total energy put into propelling a FCEB might be as low as 50% of the original energy source (i.e., the original water molecule) that went into the electrolysis process to make the hydrogen fuel in the first instance.

In this case, for each kilogram of hydrogen produced, FCEBs would require three times the amount of equivalent energy required to power a BEB over the same distance. The energy required for one FCEB (kg/km) is enough to power three BEBs (kWh/km) over the same distance. This has been factored into the financial considerations and recommendation supporting Scenario Three (mixed green fleet) being a mixed green fleet electrification solution outlined in this study.

Based on CUTRIC's modeling work and analysis, the energy consumption rate for 12 m BEBs operating in the Brampton Transit system could be in the range of 1.3 kWh/km (assuming idealized conditions without HVAC load on flat pavement). For 12 m FCEBs, the fuel consumption rate is approximately 9.1 kg/100 km. Since 50 kWh is required to produce each kilogram of hydrogen, FCEBs would incur an effective energy consumption rate of approximately 4.5 kWh/km. The energy necessary to power one FCEB is enough to power 3.5 BEBs.

Yet, there are scaled benefits associated with hydrogen fuel cell buses. At scale and for mid- to large-scale fleets (i.e., hundreds of vehicles), the overall infrastructure costs for FCEBs can be less expensive than those associated with BEBs, potentially amounting to tens of millions of dollars [91, 108].

As the world moves away from consuming fossil fuels for transportation applications, the struggle to find alternative sources for significant amounts of energy will increase the relative importance of energy efficiency to unprecedented levels. Manufacturers of FCEB technologies have acknowledged that BEBs excel in urban environments where electrical infrastructure is readily available [109]. FCEBs will likely be an essential complementary solution for the future transit, trucking and naval industries, especially for long or intense services for which BEB technology falls short of delivering the necessary range in environments or where installing charging infrastructure is overly challenging, complicated or cost prohibitive.

Notwithstanding the challenges associated with implementing FCEBs, it is interesting to note that fuel cell locomotives are already a reality in Canada. Canadian Pacific operated its first revenue kilometers using a fuel cell hydrogen locomotive in 2022 [110]. In 2023, Alstom's Coradia iLint hydrogen-powered train completed a North American demonstration in Quebec, carrying 100,000 passengers over 130 trips covering 10,660 km [111]. The Coradia iLint began its first commercial service in Germany in 2018, and has since traveled over 22,000 km in eight European countries [112]. Fuel cell technology has the potential to significantly impact the heavy-duty transportation sector more than it has to date, as demonstrated by the deployment of hydrogen-propelled rail systems in Canada recently and in Europe over the past half decade across a dozen countries.

2.3 ZEB Economy

Compared to diesel equivalents, the disadvantages associated with both battery electric and hydrogen FCEB technologies include the elevated capital costs of the vehicles (rolling stock) and associated electric charging and hydrogen fueling infrastructure required for both technologies.

However, a business case for these two decarbonization technologies can still be favorable when considering a holistic set of economic factors.

With BEB technologies, transit electrification can save fuel costs over diesel fleets. In Canadian provinces with abundant clean electrical energy, such as Ontario, charging BEBs is generally cheaper than fueling diesel buses on a route-by-route and bus-by-bus basis. While there may not be fuel savings when substituting diesel with hydrogen in Ontario today, as a result of the high cost of green (and, in some cases,

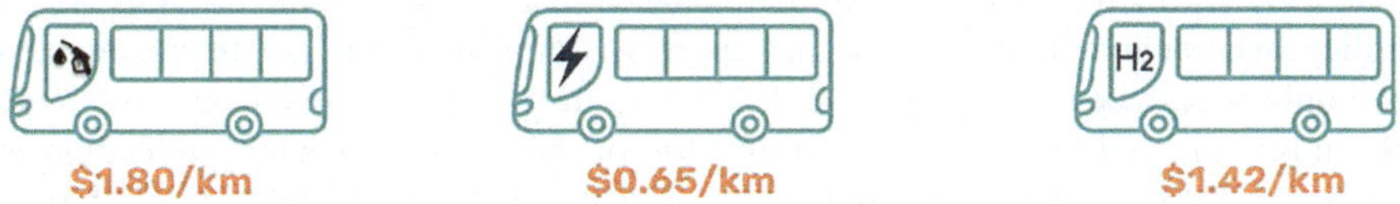

Fig. 3 National average cost per kilometer for fuel types

even for gray) hydrogen, these vehicles do reduce the social costs of smog and pollution locally.[1]

Additionally, both BEB and FCEBs are expected to allow for savings in maintenance costs over the long-term, and both technologies will save overall operational costs in any future defined by carbon pricing mechanisms at national or subnational levels by eliminating the most carbon intense fuel (diesel) used for propulsion.

Using general averages for Canada shows an average comparative diesel fuel consumption of 50 L/100 km and an average electrical energy cost of $0.35/kWh. This results in an estimated $1.80/km for diesel buses, $0.65/km for BEBs and $1.42/km for FCEBs, as illustrated in Fig. 3.

This demonstrates that even by high-level estimates, zero emissions transit technologies compete on a dollar-per-km basis operationally today. Detailed and agency-specific operational costs associated with Brampton Transit's electrification and decarbonization scenarios are discussed later in this report.

2.3.1 ZEB Maintenance Costs

Over the long term, it is expected that BEBs will have lower maintenance costs because of fewer moving parts compared to vehicles with internal combustion engines.

Early data from systems operating BEBs in the United States are encouraging. King County Metro in Washington is one of the few agencies in North America that publishes average cost of maintaining its BEBs. In 2022, the agency commissioned a test charging facility to service its new fleet of 40 BEBs, with plans to procure an additional 120 BEBs in 2025 [99, 113]. Based on empirical data from the first three buses collected from February to November 2016, King Metro reported in 2017 that the overall maintenance "down time" of BEBs is similar to that of diesel and trolleybuses and a little lower than that of hybrid buses. The availability figures in 2017 for BEBs, diesel, trolley and hybrid show 84%, 88%, 87% and 92%, respectively, in "up time" for the vehicles—meaning they are available for service. The report shows the agency has spent an average of US$0.18/mile ($0.15/km)[2] for BEB maintenance, US$0.32/mile ($0.27/km) for hybrid vehicle maintenance, US$0.44/

[1] Health care cost savings from smog and pollution reduction are not part of the extended scope of this study but could and should be analyzed in the future as part of transit decarbonization plans.

[2] 1 United States Dollar is assumed to equal 1 Canadian Dollar as of January 22, 2024.

mile ($0.37/km) for diesel vehicle maintenance, and US$0.53/mile ($0.45/km) for trolley maintenance [114]. When the report was published, the BEBs documented had only been recently deployed, and therefore the report's maintenance costs may be underestimated by the average useful life of the vehicle. A later publication in February 2018 reports higher maintenance costs of approximately US$0.26/mile ($0.22/km) for BEBs, while the other vehicle types did not experience significant changes in average maintenance costs. The latter report indicates the portion of overall costs associated with the propulsion systems is only US$0.26/mile ($0.22/km) [115]. The second value reported by King County Metro aligns with more recent assumptions regarding zero emissions bus (ZEB) fueling costs found in global literature. Quarles, Kockelman and Mohamed (2020) report global prices to be US $0.332/mile ($0.279/km) for BEBs and US$0.631/mile ($0.530/km) for diesel buses [116].

Much speculation still exists around FCEB maintenance costs. The expectation is that these costs will be similar to the maintenance costs associated with compressed natural gas (CNG) buses operating in the United States and Canada for over two decades due to their similar refueling infrastructure. It will take several years of experimental and early deployment data to generate definitive and reliable maintenance costs for FCEB fleets.

For FCEBs, potential reductions in maintenance costs due to the lack of an internal combustion engine may be offset by increased costs associated with having more systems and components onboard, such as a fuel cell stack onboard.

The National Renewable Energy Laboratory (NREL) gathered maintenance costs for 20 FCEBs in 2017. Buses running with hydrogen display a lower overall operational availability compared to their diesel bus counterparts, approximately 76% for FCEBs and 91% for diesel-running vehicles, respectively. A group of FCEBs displays total maintenance costs of US$0.56/mile ($0.47/km), while another group from a different manufacturer requires an average of US$1.25/mile ($1.05/km) for maintenance. NREL also reports that transit agencies operating FCEBs have concentrated on managing all bus maintenance in-house. These efforts resulted in transient cost increases that will no longer be incurred after those agencies go through a learning curve for operating and maintaining FCEBs [117].

Orange County Transportation Authority (OCTA) used the NREL methodology to analyze its initial 10 FCEBs for 6 months [115]. Published in March 2021, OCTA's report analyzes 883 refueling episodes in which an average of almost 19 kg of hydrogen per fueling episode is transferred. Total maintenance costs are compared with 10 CNG buses, yielding similar results. The average total maintenance costs for those FCEBs are US$0.42/mile ($0.35/km), while the average maintenance cost for CNG buses is US$0.43/mile ($0.36/km). The overall availability of FCEBs is around 73%, slightly lower than the availability observed for CNG buses at about 80% [115].

The lack of information to predict maintenance costs in the Canadian marketplace hinders a more accurate evaluation. Based on the above-cited sources, it is reasonable to expect BEBs to incur maintenance costs in the US$0.20/mile ($0.17/km) to US$0.30/mile ($0.25/km) range, and FCEBs will incur maintenance costs in the US

\$0.40/mile (\$0.36/km) to US\$0.60/mile (\$0.50/km) range. Based on these results, maintenance costs in Canada are expected to range between \$0.17/km and \$0.25/km for BEBs and between \$0.33/km and \$0.50/km for FCEBs. Across all NREL studies reviewed, most of the maintenance is not associated with the powertrain issues but with general bus issues, which explains the lower maintenance cost for ZEBs than diesel vehicles.

2.3.2 Hydrogen Price Points

Hydrogen fuel prices in Ontario will be affected by global and local marketplace conditions in the future, as the fuel is transportable and generated by globalized energy companies and local utilities.

Currently, Bloomberg has evaluated price point ranges for the production of green hydrogen to be between US\$4.50/km (\$3.79/km) and US\$12/kg (\$16.2/kg), blue hydrogen between US\$1.80/kg (\$2.43/kg) and US\$4.70/kg (\$6.35/kg), and gray hydrogen between US\$0.98/kg (\$1.32/kg) and US\$2.93/kg (\$3.96/kg) (in 2023 dollars) [118]. This study predicts that green hydrogen price ranges will match predicted blue hydrogen price ranges by 2030 and will become slightly cheaper by 2050 [104]. However, these figures do not include the cost of delivering the fuel, which causes a significant increase in fuel cost as the delivery entails expensive dispensers to fuel vehicles along with the hydrogen itself. These costs also fail to incorporate unknown costs associated with carbon capture and storage technology, which may affect the long-term viability and price point for blue hydrogen ("blue" hydrogen remains a hypothetical fuel in Canada due to a lack of functional carbon capture and storage facilities that mitigate carbon emissions).

Domestically, Natural Resources Canada (NRCan) has evaluated hydrogen production costs after considering the results of several studies abroad and two Canadian case studies. The results are aligned with those reported by Bloomberg on production costs. However, NRCan estimates the costs of delivering hydrogen fuel vary substantially given differing scenarios of distance covered and the complexities of transportation of gaseous and liquid hydrogen. NRCan reports that the price point for the delivered hydrogen ranges from \$5/kg[3] to \$12/kg [104]. Still, according to NRCan, the projections for 2050 are in the range of \$1/kg to \$3.5/kg, even for the delivered hydrogen fuel [104].

The Hydrogen Council, a global non-profit representing industry groups in the hydrogen industry, has also offered estimates of the current average price points for blue and green European hydrogen delivered through different methods. The total cost of gaseous trucking is about US\$11/kg (\$11/kg[4]), whereas liquid trucking is about US\$10/kg (\$10/kg) [119]. Through a pipeline, the total cost is estimated to be

[3]Unless otherwise noted, all funds are in Canadian dollars.

[4]This conversion used the average rate for United States Dollar to Canadian Dollar average for November 2012 (the time of the study) at US\$1 = \$0.9970.

around US$12/kg ($12/kg) with a decrease of about 60% expected by 2030 for all delivery methods [102].

Ballard Power Systems Inc. estimates the price point of delivering hydrogen using liquid storage or local SMR production is in the range of US$3/kg ($4.0/kg[5]) to US$6/kg ($8.0/kg) if a commercial project with more than 20 buses is considered [120]. The company estimates that the range of US$10/kg ($13.4/kg) to US$12/kg ($16.0/kg) would more accurately represent delivered fuel costs for demonstration fleets with a size of five buses [106].

Given these variations in estimated and predicted hydrogen price points in Canada, a range of $8/kg to $40/kg is used in this study based on results from a market scan carried out in 2023 for CUTRIC's Pan-Canadian Hydrogen Fuel Cell Electric Bus Demonstration and Integration Trial [121]. While it is not easy to estimate the capital costs of hydrogen generation and refueling infrastructure, as the characteristics of each project determine those costs precisely, Ballard Power Systems proposes that the infrastructure required to perform on-site SMR or liquid storage handling up to 1000 kg daily would include a vaporizer, a cryo-pump, storage tanks and dispensers. This would cost between US$3.5 million ($4.7 million[6]) and US$8.5 million ($11.4 million) [106, 120].

Thus, the cost analyses affecting Scenario Two (full FCEB solution) rely on the price points of $25/kg, $20/kg, and $8/kg for green, blue and gray hydrogen, respectively, in the economic analysis presented later in this study. An additional economic case is included using an elevated potential green hydrogen price point of $40/kg to capture high prices due to market volatility. The hydrogen market in Ontario is further discussed in the next section of this report.

2.3.3 Ontario's Hydrogen Technology Pricing and Landscape

This section sheds light on the dynamic landscape of hydrogen technology in Ontario's public transit sector. The Province of Ontario's proactive commitment to sustainable hydrogen production, facilitated through various production methods and recent investments in hydrogen infrastructure across the province, signifies a strategic initiative that will positively affect the pursuit of the public transit agency's decarbonization strategy, as described later in this report.

Based on the latest available data, the cost of hydrogen for fuel cell buses in Canada varies significantly, with prices observed up to $60/kg in market scans completed for CUTRIC's Pan-Canadian Hydrogen Fuel Cell Electric Bus Demonstration and Integration Trial [121]. This wide range is attributed to diverse

[5]This conversion used the average rate for United States Dollar to Canadian Dollar average for March 2019 (the time of the information was obtained) at US$1 = $1.3368 Canadian.

[6]This conversion used the average rate for United States Dollar to Canadian Dollar average for March 2019 (the time of the information was obtained) at US$1 = $1.3368 Canadian.

production methods, energy inputs and technological advancements that impact the overall cost structure of locally produced and delivered hydrogen supplies.

On average, hydrogen costs are significantly higher compared to diesel costs across Canada due to limited operational supplies and increased demand for low-carbon hydrogen supplies. As more facilities come online and the market becomes more competitive, hydrogen costs are anticipated to reduce over time. However, variable supply chain issues and unpredictable national and global inflationary pressures may continue to press prices higher (rather than lower) over the next several years.

In Ontario, a strategic commitment to developing a sustainable hydrogen sector is evident through the adoption of various production methods and investments made with provincial funding in recent years. The province is exploring hydrogen production through electrolysis, steam methane reformation (SMR) and biomass gasification, which leverage organic materials for hydrogen production, emphasizing minimizing the carbon footprint. This array of methods highlights the economic dynamics specific to Ontario's hydrogen market, where decisions balancing cost-effectiveness and environmental impact will significantly influence the province's energy landscape.

At the forefront of its commitment to environmental sustainability and energy innovation is *Ontario's Low-Carbon Hydrogen Strategy*, designed to propel the province towards a cleaner and more efficient energy future [122]. This strategic initiative outlines a framework encompassing research, development, and deployment of low-carbon hydrogen technologies, laying the foundation for a robust hydrogen economy.

Ontario's hydrogen strategy envisions hydrogen as playing a key role in provincial economic growth and energy cost reductions. The province aims to leverage surplus electricity from wind and nuclear sources to produce low-carbon hydrogen, as the planned Niagara Hydrogen Centre demonstrated. The Independent Electricity System Operator (IESO) also identifies hydrogen as a cost-effective resource for reducing peak demand, with projections indicating substantial hydrogen-generating capacity by 2050 [123].

The province has committed to advancing hydrogen technologies through proactive policies and initiatives. The Hydrogen Innovation Fund stands out as a key pillar with a $15 million investment over 3 years that signifies Ontario's commitment to integrating hydrogen into its clean electricity system. This strategy aims to stimulate advancements in hydrogen-related projects. This fund supports research, development, and commercialization efforts in the hydrogen sector, fostering innovation and contributing to Ontario's broader clean energy goals.

Ontario's Ministry of Energy is considering several options to provide reduced electricity rates for hydrogen producers to promote sustained growth in Ontario. It introduced a proposal under the ERO number 019-5381 aligned with Ontario's Low-Carbon Hydrogen Strategy, which elicited public feedback during a 60-day comment period that ended on June 13, 2022 [124]. In February 2023, Ontario created the Hydrogen Innovation Fund as part of its Low-Carbon Hydrogen Strategy. The Hydrogen Innovation Fund was established to invest C$15 million over

3 years to initiate and develop opportunities for hydrogen integration into Ontario's electricity system, including hydrogen electricity storage [125]. The proposal outlines three distinctive options to encourage the expansion of Ontario's low-carbon hydrogen economy. The first option amends the Industrial Conservation Initiative (ICI) rules to broaden eligibility for hydrogen producers. The second option envisions businesses co-locating hydrogen electrolyzers at electricity generation facilities using curtailed generation. The third option proposes the development of an interruptible rate pilot specifically tailored for hydrogen producers. These initiatives will take years to take effect, so their effects on hydrogen pricing today are minimal. This is expected to change over the next 5 years in favor of lower pricing for green hydrogen supplies as Ontario's current hydrogen production investments take hold and impact the production volume.

There are signs of positive advancement in this regard. Across Ontario today, hydrogen production projects are proposed in municipalities including London, Ottawa, Toronto, Markham, Hamilton and Niagara Falls. Notably, a federally funded Enbridge and Cummins collaboration that resulted in the Markham Energy Storage Facility being built in the past few years remains the first utility-scale facility in North America, blending hydrogen into the natural gas system.

In tandem, since 2017, CUTRIC has worked to get the first large-scale FCEB demonstration project launched in Ontario to showcase the technology, learn from deployment data and motivate the creation of a sustainable hydrogen supply chain specifically for public transit. CUTRIC's Pan-Canadian Hydrogen Fuel Cell Electric Bus Demonstration and Integration Trial project is at the forefront of this effort, with the City of Mississauga and MiWay as the lead public transit agency and champion. The FCEB project in Mississauga seeks to showcase an integrated green hydrogen supply chain while actively promoting the operational reliability, cost-effectiveness and environmental advantages of FCEBs as a sustainable solution. CUTRIC has established significant partnerships with industry leaders and transit agencies to encourage collaboration across the project partnership to expedite the development and deployment of hydrogen fuel cell technology in Canada's public transit industry.

One significant concern delaying the launch of the MiWay FCEB project is the high upfront costs associated with establishing a comprehensive green hydrogen infrastructure for the transit agency. Ballard Power Systems has released information indicating that the cost of hydrogen fuel cell vehicles is expected to equalize with diesel around 2028, which may help reduce some of the costs of transitioning to hydrogen-powered decarbonized fleets (see Sect. 2.3.2 above regarding predicted hydrogen fuel ranges).

The future growth and opportunities associated with advancing hydrogen fuel cell technology within Ontario's public transit sector hold immense promise for sustainable zero emissions transportation. As the province seeks to reduce carbon emissions and combat climate change, hydrogen fuel cell buses and heavy-duty vehicles will emerge as a viable solution, offering a clean and efficient alternative to traditional fossil fuels. Ontario's public transit sector benefits significantly from adopting this cutting-edge technology, paving the way for a greener and more sustainable future.

2.4 Case Studies of Municipalities Pursuing Decarbonization

According to the *Canadian ZEB Database™ Report* (2023), a CUTRIC study of 56 Canadian transit agencies, 5426 ZEBs are at different stages of transition, starting from "pronouncements" through "feasibility studies," "funding," "procurement," "commissioning" and "in-service" at municipal and regional transit agencies nation-wide. BEBs constitute 5330 ZEBs, while FCEBs constitute 96 of the vehicles tallied [67].

As of October 2023, a total of 2796 BEBs are in the pronouncement stage, 432 BEBs are undergoing feasibility or have completed their feasibility studies recently, 566 BEBs are in procurement mode and another 231 BEBs have been commissioned with riders onboard. There are 45 FCEBs in the pronouncement stage, 50 FCEBs undergoing feasibility studies, 10 FCEBs in funding stages, 29 FCEBs under procurement and three buses in the commissioning stage. Two FCEBs are currently in the pilot deployment stages.

2.4.1 Toronto Transit Commission (TTC): Toronto, Ontario

The Toronto Transit Commission (TTC) has established a goal of achieving a 50% zero emissions fleet between 2028 and 2032 and a fully zero emissions fleet by 2040. The agency is taking significant steps toward achieving this goal by electrifying its entire fleet. As part of the TTC Green Initiatives, the TTC has one of North America's largest fleets of electric buses, with 60 vehicles in operation as of 2019 [53]. In addition, the TTC has 1162 buses under procurement from New Flyer and Nova Bus.

2.4.2 TransLink: Vancouver, British Columbia

With a long-term goal of introducing 400 new BEBs to its fleet by 2030 and fully transitioning the fleet to zero emissions vehicles by 2040, TransLink received its first 15 BEBs in November 2022 [126]. TransLink's Climate Action Plan includes expanding the CNG bus fleet to replace 84 diesel buses by the end of 2024, replacing 34% of the diesel fleet with 462 BEBs, and redeveloping and building new charging infrastructure by 2030 to serve its electric fleet [127, 128]. TransLink currently has 155 buses undergoing feasibility studies.

2.4.3 Société de Transport de Montréal (STM): Montreal, Quebec

The *Société de transport de Montréal* (STM) had its first fully electric bus hit the road in 2017, and in 2019, STM received 30 long-range BEBs. This increase marked the first step in the agency's goal to acquire only electric buses from 2025 to reach zero GHG emissions by 2040 [129]. In February 2020, the STM's 36 Monk route became Canada's first fully electric bus transit line. Currently, the agency fleet has a mixed fleet of 42 electric buses, including fast-charging and depot-charged buses, as well as midibus and minibus [130].

2.4.4 Edmonton Transit Services (ETS): Edmonton, Alberta

Edmonton Transit Service (ETS) is the first Canadian transit agency to purchase many BEBs in one procurement round, obtaining 50 Proterra buses in 2018 [91]. The City of Edmonton announced its plan in 2021 to become carbon neutral by 2050. ETS has operated 60 BEBs since 2020, with two FCEBs unveiled for pilot use in September 2022. ETS has initiated planning and action toward complete electrification of its fleet [65].

2.4.5 Kingston Transit Services: Kingston, Ontario

As part of the City of Kingston's Strategic Plan 2019–2022, the city aims to electrify its fleet of public transit vehicles, incrementally. This objective comes with a key performance indicator of acquiring 12 electric buses by 2022 [131]. Kingston Transit Service's first two electric buses entered service in July 2021. The City of Kingston's 2022 approved capital budget included the purchase of five additional electric buses, which were slated for delivery in 2023. In July 2022, the City of Kingston also secured funding to conduct a full fleet feasibility study and transition plan for its transit bus fleet [132].

3 Summary Analysis

As noted above, zero emissions buses (ZEBs) present several advantages over conventional diesel buses. However, ZEBs are complex energy systems requiring high upfront investments in supporting charging and/or new fueling systems, and their actual maintenance costs over the long term are unknown. The results of the analyses above are summarized below, indicating that the "best" balanced option ahead for a city and transit agency like that of Brampton (mid-sized agency) is to pursue a mixed fleet solution by adopting and deploying both BEBs and FCEBs over a 20+ year implementation period.

Scenario One (Full BEB Solution)

In this scenario, the study assumes only battery electrified buses are used to achieve decarbonization. This scenario analyzes two charging strategies: "Charging Strategy One: depot only charging solution" and "Charging Strategy Two: depot with on-route charging." In Charging Strategy Two (depot with on-route charging), on-route charging is assumed only to ensure the bus can complete its daily duties while maintaining the required power levels. Modeling results are optimized to reduce the need for on-route charging, given the high cost and complexity of high-powered charging systems, which require extensive civil works.

This scenario shows that the decarbonization of the current service with BEBs exclusively will require between 570 and 674 buses, depending on the vehicle manufacturer and cabin heating configuration. This amounts to 20–42% more vehicles than the base case fleet size. With the current fleet split, the fleet would be composed of 462 BEBs that are 12 m long and 108 BEBs that are 18 m long buses or (depending on the specific platforms deployed on vehicles) 456 BEBs that are 12 m long and 128 BEBs that are 18 m long. The service hours required for the current fleet would increase by approximately 9%. This transition will require facility retrofits and installation of approximately 22 MW of charging capacity between the two Brampton Transit facilities, and 18 on-route chargers. Therefore, the on-route chargers required will amount to 600% more than the four chargers the transit agency has deployed in the base case. These chargers will need to be deployed across eight terminals.

Factoring in expected service growth figures, the public transit agency would require a fleet of between 1161 and 1240 BEBs, the consumption of between 106 GWh and 131 GWh of energy per year, and the completion of a third transit facility with additional charging power capacity of approximately 16 MW by 2040.

Scenario Two (Full FCEB Solution)

In this scenario, the study assumes only hydrogen-fueled electrified buses (FCEBs) are used to achieve decarbonization. It also assumes that FCEBs can be fueled with various production types of hydrogen, as defined below:

1. Electrolytic or "green" hydrogen produced from the electricity grid is considered a low-carbon-intensity hydrogen fuel type.
2. "Blue" hydrogen is produced from steam methane reformation (SMR), which integrates carbon capture solutions to store emissions and reduce the overall carbon intensity of the fuel. It is considered a greener solution compared to "gray" hydrogen defined below but a dirtier, more carbon intense solution than "green" hydrogen.
3. "Gray" hydrogen is produced from steam methane reformation (SMR) with no mitigating technology to reduce the carbon intensity of the fuel. This production method is considered the dirtiest form of hydrogen, but it is abundantly available at scale within Canada today.

This scenario shows that the decarbonization of the current service using FCEBs exclusively will require 720 buses in total (51% more than the base case fleet size), but with no depot or on-route chargers. The fleet would be comprised of 583 FCEBs that are 12 m in length and 137 FCEBS that are 18 m in length. The service hours to maintain current service would only increase by 2%. It will, however, require new hydrogen supply agreements with vendors, which may not provide the required volumes of green hydrogen necessary for several years, requiring the use of more carbon-intense gray options in the short term.

Factoring in the expected service growth, the public transit agency would require a fleet of approximately 1418 FCEBs, the completion of a third transit facility equipped to receive hydrogen supplies and the procurement of 5.9 million kilograms (kg) of hydrogen fuel per year by 2040.

Scenario Three (Mixed Green Fleet Solution)

In this scenario, the study aims for an optimized combination of BEBs and FCEBs, reducing the number of on-route chargers and charging episodes for BEBs and the total number of buses for full-fleet decarbonization.

This scenario shows that decarbonization of the current service with a mixed green fleet solution composed of BEBs and FCEBs will require between 558 and 640 buses in total (a 17–34% increase in vehicles compared to the base case), as well as facility retrofits and installation of approximately 8 MW of charging capacity. Despite optimization efforts to reduce the number of chargers required, this scenario still requires the same 18 on-route chargers as Scenario One (full BEB solution). To maintain current service, service hours would increase by approximately 9%. Scenario Three (mixed green fleet solution) would require new hydrogen supply agreements with vendors that may not be able to provide the necessary volumes of green hydrogen for several years. Using more carbon-intense gray hydrogen options in the short term may be necessary.

The comparative analysis of all three scenarios shows that a mixed green fleet solution would comprise 318–413 BEBs and 145–322 FCEBs. In one configuration, the fleet would comprise 258 BEBs that are 12 m long, 60 BEBs that are 18 m long, 117 FCEBs that are 12 m long and 28 FCEBS that are 18 m long. In the other configuration, the fleet would comprise 335 BEBs that are 12 m long, 78 BEBs that are 18 m long, 261 FCEBs that are 12 m long and 61 FCEBs that are 18 m long.

This achieves the goal of the smallest total vehicle fleet size to decarbonize the current fleet of the public transit agency's buses. However, this small size in fleet vehicles does not equate to minimal total costs or maximum potential emissions reduction.

In addition to the expected service growth, the public transit agency would require by 2041 a fleet of between 724 and 767 BEBs and between 408 and 418 FCEBs. Furthermore, a third transit facility with an additional charging capacity of approximately 16 MW would need to be completed. The annual electrical energy consumption would increase to between 106 and 131 GWh procuring approximately 1.3 to 2.2 million kg of hydrogen fuel annually.

Table 2 summarizes the ZEB fleet required for each scenario.

Charging System Impacts on Infrastructure

This study also explores the results of optimizing the public transit agency's fleet and charging infrastructure. The results of the optimization work carried out by CUTRIC demonstrate the following impacts on service hours when BEBs, FCEBs and/or a combination of the two technologies are leveraged to decarbonize a public transit fleet:

Table 2 Required ZEB fleet size

		Base case		Scenario One (full BEB solution)		Scenario Two (full FCEB solution)	Scenario Three (mixed green fleet solution)			
							Configuration one		Configuration two	
		Diesel/ HEB	BEB	Configuration one Types 1 & 3	Configuration two Type 2	Types 4 & 5	Types 1 & 3	Types 4 & 5	Type 2	Types 4 & 5
Current service	BEBs without DFAH	468	8	633	674	720	355	247	318	322
							Total = 602		Total = 640	
	BEBs with DFAH			590	570		394	181	413	145
							Total = 575		Total = 558	
Growth fleet (2041)	BEBs without DFAH	920	18	1240	–	1418	767	418	–	–
							Total = 1185		–	
	BEBs with DFAH			1161	–		724	408	–	–
							Total = 1132		–	

- Scenario One (full BEB solution) and Scenario Three (mixed green fleet solution) have similar impacts on service hours with an increase of between 2% and 9% with diesel auxiliary heaters (depending on vehicle configuration) and an increase of between eight and 11% without self-contained clean-diesel-fired heaters (DFAH).
- Scenario Two (full FCEB solution) has the least impact on service hours with only a 2% increase over the base case. This reduced impact is because FCEBs operate similarly to diesel buses from a fueling standpoint and do not require charging during operations.

This study further assessed the worst-case scenario, i.e., the "heavy-duty cycle" scenario, across the entire fleet to determine what the maximum impact on capital cost, infrastructure and service hours would be using BEBs, FCEBs or a combination of the two. Based on this assumption, the public transit agency would need to procure dozens of new BEBs or FCEBs, or a combination of the two technologies each year to achieve full-fleet electrification by 2040.

Greenhouse Gas (GHG) Emissions

From a greenhouse gas (GHG) perspective, considering all emissions ranging from the manufacturing of assets to facility retrofits and bus operations as part of an environmental life cycle analysis (LCA) of GHG, three scenarios emerge.

For the base case scenario, which would see the public transit agency continue operating diesel-powered buses, the agency would grow from 476 to 920 vehicles, and its operational GHG emissions would grow from approximately 55.32 $ktCO_2$/year to approximately 108.80 $ktCO_2$/year.

For Scenario One (full BEB solution), depending on vehicle specifications and cabin heating configurations, after a ramp-up of approximately 12 years procuring an increasing number of buses, the public transit agency will need to procure approximately 81 BEBs per year until approximately 2041. Afterwards, the procurement volumes start to reduce as fewer diesel buses need replacement and procurement is more directed to address service growth and replace BEBs. This scenario retains the public transit agency's eight BEBs and four high-powered chargers from past deployments. In this scenario, the public transit agency would decrease operational GHG emissions from approximately 55.32 $ktCO_2$/year to approximately 2.78 $ktCO_2$/year.

For Scenario Two (full FCEB solution), after a ramp-up of 11 years procuring an increasing number of buses, the public transit agency will need to procure 92 FCEBs per year until approximately 2041. Afterwards, the procurement volumes start to reduce as fewer diesel buses need replacement, while procurement is directed to address service growth and replace FCEBs. The operation of a full FCEB fleet would emit between 16.34 $ktCO_2$/year and 80.71 $ktCO_2$/year depending on the type of hydrogen fuel used.

For Scenario Three (mixed green fleet solution), depending on the BEB manufacturer and its cabin heating configuration, after a ramp up approximately 8 years of procuring an increasing number of buses, the public transit agency will need to procure approximately 81 BEBs per year and approximately 62 FCEBs per year until approximately 2041. Afterwards, procurement volumes start to reduce as fewer diesel buses need replacement, and procurement is directed to address service growth and replace ZEBs. This scenario retains the eight BEBs and four high-powered chargers the city already possesses from past deployments. In this scenario, the operational GHG emissions would be between 5.16 ktCO_2/year and 24.53 ktCO_2/year by 2041, depending on the type of hydrogen fuel used.

These scenarios are aggressive and presuppose full facility modifications commence and are completed by the end of 2025.

Overall, these assessments show all three electrification scenarios achieve substantial emissions reductions over the diesel base case scenario. These emissions reductions would aid the city's of Brampton's goals to reduce the total GHG emissions generated by 80% by 2050.

Scenario One (full BEB solution) is a highly effective means of reducing GHG emissions across Brampton's public transit fleet. It is also the most environmentally friendly option of the three decarbonization scenarios available, as the environmental LCA shows that BEBs could save approximately 69% of all emissions compared to the base case.

Scenario Two (full FCEB solution) is the least effective means of reducing GHG emissions. The environmental LCA shows that FCEBs could emit between 69% less and 25% more than diesel buses, depending on the hydrogen fuel type and bus diesel configuration.

Scenario Three (mixed green fleet solution) presents an intermediary effectiveness in reducing GHG emissions. The environmental LCA shows this solution could save between 25% and 61% of all emissions, depending on the cabin heating configuration of BEBs and the method of hydrogen fuel production for FCEBs.

These relative LCA emissions reductions are expected to be independent of service growth but could change substantially if the carbon intensity of the electricity grid or hydrogen fuel supply chain changes in the future.

Considering only the operational GHG emissions of the current service levels, in Scenario One (full BEB solution), Brampton Transit could eliminate between 53.19 and 54.37 ktCO_2eq/year[7] if BEBs were equipped with DFAH or rely solely on the battery for heating, respectively. These figures are tabulated using the public transit agency's total fuel consumption as of 2019 and the respective carbon intensity factors associated with diesel fuel (factoring in the agency's usage of biodiesel), compared to the total fuel/energy consumption and respective carbon intensity factors for electricity and various types of hydrogen. Avoiding the emission of 54.37 ktCO_2eq/year could constitute relative reductions of 98% in operational emissions for Brampton Transit. This reduction is the equivalent of removing more than 12,000 passenger cars from city roads,[8] depending on the cabin heating

[7] Tonnes of carbon dioxide equivalent per year

[8] Assuming that the typical Canadian passenger car consumes 8.6 L/100 km and travels approximately 15,200 km/year, considering that the energy density of gasoline is 46 MJ/kg and the volumetric density of gasoline is 0.743 kg/L. Therefore, the average Canadian cars emit approximately 4.24 tCO2/year.

configuration [3, 4]. Factoring in the expected growth of the public transit agency services, the absolute operational GHG emission reductions by 2040 would be approximately 106.02 ktCO_2eq/year equivalent to removing approximately 23,338 passenger cars from the roads annually.

The comparative operational emissions savings resulting from Scenario Two (full FCEB solution) or Scenario Three (mixed green fleet solution) depend upon the production method used for the hydrogen. In Scenario Two (full FCEB solution), the public transit agency could eliminate anywhere from approximately 50.27 ktCO_2eq/year (~90%) using green hydrogen supplies to 37.67 ktCO_2eq/year (~68%) using blue hydrogen[9] supplies and 19.66 ktCO_2eq/year (~35%) using gray hydrogen supplies, considering current service levels. Factoring in the expected growth of the public transit agency services, the emissions savings for the total fleet could be up to 92.46 ktCO_2eq/year using green hydrogen supplies and 28.10 ktCO_2eq/year using gray hydrogen supplies by 2040.

In Scenario Three (mixed green fleet solution), considering current service levels, the public transit agency could eliminate anywhere from 42.59 ktCO_2eq/year (~77%) using gray hydrogen supplies for FCEBs and by equipping BEBs with DFAH to 53.09 ktCO_2eq/year (~95%) using green hydrogen supplies for FCEBs and relying solely on batteries to heat BEB cabins. Factoring in the expected growth of the public transit agency services, the operational GHG emissions savings can be anywhere from 84.27 tCO_2eq/year (~77%) using grey hydrogen supplies for FCEBs and by equipping BEBs with DFAH to 103.64 tCO_2eq/year (~95%) using green hydrogen supplies for FCEBs and by relying solely on batteries to heat BEB cabins by 2040.

Economic Cost Analysis

The study also provides the full economic cost of the three full-fleet decarbonization scenarios by assessing the total cost of ownership (TCO) for 12 m and 18 m buses across all three scenarios. The analysis compares those costs to a baseline economic analysis of the public transit agency's current diesel fleet. TCO is defined as the cost of purchasing and operating the vehicle (or fleet) over the entire lifespan of the vehicle. This section includes the rolling stock procurement costs and maintenance and fuel costs, among other variables that comprise the full cost of owning and operating a given vehicle or fleet type.

Two bus life cycle scenarios are considered—15 years and 18 years. Over a 15-year life cycle, the TCO for a 12 m conventional diesel bus is $1.77 million in 2023 dollars (the "base case" comparison for Brampton). The calculation for a 12 m diesel electric-hybrid bus (HEB) shows a cost of $2.09 million (NPV). In contrast, Scenario One (full BEB solution) costs $2.95 million (NPV) per BEB if those vehicles rely solely on batteries for heating and are equipped with DFAH, respectively. Scenario Two (full FCEB solution) costs from $3.99 million (NPV) to $4.80 million (NPV) per FCEB, depending on whether the agency uses lower priced gray

[9]The production of blue hydrogen in Canada remains hypothetical.

hydrogen or more expensive green hydrogen. This analysis shows that the TCO for an 18 m diesel-electric hybrid bus is $2.34 million in 2023 dollars, compared to $4.96 million in Scenario One (full BEB solution), depending on the cabin heating configuration. The TCO in Scenario Two (full FCEB solution) is between $5.76 million and $7.62 million, depending on the hydrogen fuel production method. The TCO for a 12 m or 18 m BEB or FCEB in Scenario Three (mixed green fleet solution) stays relatively stable per bus level. The same analysis was done for an 18-year life cycle and can be found in Section Eight.

The analysis shows the cost of full-fleet decarbonization across all three scenarios. Over a 15-year life cycle, the base case scenario with diesel buses costs approximately $7.92 billion (NPV). Scenario One (full BEB solution) costs between $8.96 billion (NPV) and $9.21 billion (NPV). Scenario Two (full FCEB solution) costs between $9.74 billion (NPV) and $10.12 billion (NPV) depending on method of hydrogen production. Finally, Scenario Three (mixed green fleet solution) costs between $9.00 billion (NPV) and $9.23 billion (NPV) depending on BEB heating and hydrogen production. For an 18-year life cycle, the base case scenario with diesel buses costs $7.56 billion (NPV). Scenario One (full BEB solution) costs between $8.96 billion (NPV) and $9.20 billion (NPV). Scenario Two (full FCEB solution) costs between $9.84 billion (NPV) and $10.22 billion depending on hydrogen production method. Finally, Scenario Three (mixed green fleet solution) costs between $8.94 billion (NPV) and $9.18 billion (NPV), depending on the cabin heating configuration and hydrogen production method.

The economic analysis also includes the cash flow, or future dollar value (FDV), analysis of the transition scenarios. The cash flow refers to the movement of money into and out of a business's accounts over a specific period of time. It represents the cash generated or consumed by a company's operations and investments. Reviewing the three scenarios with a 15-year useful life and no DFAH, the total costs would show the base at $11.42 billion (FDV), Scenario One (full BEB solution) at $13.29, Scenario Two (full FCEB solution) at $14.73 billion (FDV) and Scenario Three (mixed green fleet solution) at 13.23.

Within the economic analysis, a detailed facility assessment determines whether existing transit facilities can support a fleet-wide transition to BEBs and FCEBs from a civil, architectural, electrical, mechanical, structural and technical standpoint. The outcomes of this analysis show that the total cost of the retrofits, including equipment purchase and installation, is expected to be approximately $104 million (NPV) for the agency's first facility, and $55 million (NPV) for its second facility. Depending on the number of chargers at each site, the on-route charger locations are expected to cost between $700,000 (NPV) and $2.7 million (NPV) each.

Implementation Plan

Based on the study carried out for this report, an effective implementation plan is one that would guide the transit agency in determining the impact of BEB versus FCEB bus procurements and deployments.

For Scenario One (full BEB solution), after approximately 11 years of ramping up procurement, the public transit agency will need to procure approximately 81 BEBs annually until approximately 2040. Afterwards, the procurement volumes start to reduce as fewer diesel buses need replacement, and procurement is directed to address service growth and replace BEBs.

For Scenario Two (full FCEB solution), the public transit agency will need to procure 92 FCEBs per year until approximately 2040. Afterwards, the procurement volumes start to reduce as fewer diesel buses need replacement, and procurement is more directed to address service growth and replace FCEBs.

For Scenario Three (mixed green fleet solution), the public transit agency will need to procure approximately 86 BEBs annually in the early years of transition and approximately 68 FCEBs annually in the later years until 2040. Afterwards, the procurement volumes start to reduce as fewer diesel buses need replacement, and procurement is directed to address service growth and replace ZEBs.

3.1 Conclusion

The present study results in the following conclusions based on a deep dive analysis of BEB versus FCEB technologies in a Canadian public transit application.

1. *Scenario selection*

All three scenarios—Scenario One (full BEB solution), Scenario Two (full FCEB solution) and Scenario Three (mixed green fleet solution)—come with advantages and disadvantages.

Scenario One (Full BEB Solution): Key Advantages
A full BEB implementation could be an easy pathway to decarbonization because it necessitates installing and handling a readily available electricity fuel supply chain. Due to the agency's early participation in ZEB pilot adoption in previous years, the public transit agency is already familiar with this technology.

This solution would require significant electrical infrastructure upgrades (both downstream and upstream from the facilities' transformers). However, the required electrical infrastructure is readily available in the marketplace today.

In addition, the GHG emissions associated with the operation of BEBs are shown to be less than those of FCEBs, even when considering the cleanest form of hydrogen (electrolytic or green hydrogen). A full environmental life cycle analysis that factors in the vehicle manufacturing and associated infrastructure confirms that BEBs are cleaner than FCEBs over their respective life cycles. For this reason, a full BEB implementation would also realize the greatest GHG savings for Brampton Transit.

Scenario One (Full BEB Solution): Key Disadvantages
A full BEB solution has the disadvantage of requiring substantial growth in the fleet size with hundreds of additional buses necessary and associated service hours to achieve full decarbonization over the base case scenario. This pathway constitutes a growth in vehicles of between 20 and 42%. Such high replacement ratios would

create demands for more space for buses to serve the same ridership population as today.

A BEB-only implementation solution also requires relatively large electrical infrastructure—specifically, 22 MW of charging capacity distributed between the two facilities that the public transit agency currently operates and 18 on-route chargers distributed among eight transit terminals.

Lastly, this solution does not consider unknown factors of complexity that may arise with a BEB-only solution, such as a system-wide power outage.

Scenario Two (Full FCEB Solution): Key Advantages
A full FCEB solution has the advantage of requiring the smallest increase in service hours while achieving full fleet decarbonization.

Relying solely on FCEBs for full fleet electrification does not require the substantial electrical infrastructure associated with charging BEBs. This advantage has a great potential to simplify large fleet electrification projects by eliminating the necessity of on-route charging. The public transit agency would not need to purchase, lease, or engage in land use agreements for pieces of land required to install the necessary electrical infrastructure, which is either private property or located in other municipalities.

Refueling FCEBs is operationally similar to refueling diesel buses. This facilitates the adoption of the new technology by transit staff.

Scenario Two (Full FCEB Solution): Key Disadvantages
Although FCEBs have longer ranges compared to BEBs that charge only in the depot, BEBs charging both in-depot and on-route can have a much longer range than FCEBs, which have access to only one fueling episode per day. Consequently, a full FCEB solution produces the largest number of vehicles to achieve full decarbonization.

This solution also requires the continuous procurement of large quantities of hydrogen fuel, which is expensive in Ontario today in "green" forms. Currently, the green hydrogen marketplace is virtually non-existent. The public transit agency may have to pay more than $20 per kilogram for green hydrogen, which increases operational costs significantly (although gray hydrogen may be procured at a lower cost). FCEBs are also approximately 20% more expensive than BEBs in today's marketplace.

Scenario Three (Mixed Green Fleet Solution): Key Advantages
Scenario Three (mixed green fleet solution) using BEBs equipped with DFAH is the least costly solution for the public transit agency, being cheaper than Scenario One (full BEB solution) at $9.21 billion (NPV) and more affordable than Scenario Two (full FCEB solution) at $9.74 billion (NPV) over the project timeline to 2040. Scenario Three (mixed green fleet solution) would cost $9.17 billion (NPV).

A mixed green fleet solution that employs both BEBs and FCEBs also has the advantage of integrating both technologies for fleet flexibility.

This solution also reduces the number of vehicles and chargers needed to achieve full decarbonization, significantly reducing the challenges associated with storage space in the two transit facilities.

The mixed green fleet scenario offers a compromise between less costly buses with BEBs and the smallest increase in added service hours with FCEBs.

Scenario Three (Mixed Green Fleet Solution): Key Disadvantages
Implementing two new technologies could double the challenge for the public transit agency. The agency would need to maintain two types of buses, maintain two parts inventories, develop two training programs in workforce development, manage two types of fueling infrastructure, obey two sets of standards and safety regulations, and constantly procure two fuel types. It would also have similar service hours increase as Scenario One (full BEB solution) with approximately 9% increase.

The outcomes of this holistic study—including both strategic and technical considerations discussed in *Volume One* and *Volume Two*, respectively—support a midway approach to full fleet decarbonization that prioritizes Scenario Three (mixed green fleet solution). This solution balances the advantages and disadvantages of BEB and FCEB technologies for deploying a flexible future fleet.

Although this solution requires hydrogen supplies in the near-term future, which may not be readily available in electrolytic or green hydrogen forms, the public transit agency can access gray hydrogen supplies at scale within the marketplace. This gray hydrogen supply chain would be a part of the early-stage transition plan to achieve net zero emissions, while waiting for green supplies to enter the market at greater scale and reduced cost.

Scenario One (full BEB solution) is positioned as the second-best option for the public transit agency, as it portrays a mid-way cost among the three scenarios, although it carries more significant benefits regarding emissions savings. Scenario Two (full FCEB solution) is the third best option given the cost profiles associated with FCEBs in today's marketplace in Canada and the uncertain green hydrogen supply chain in Ontario that impacts total cost and total greenhouse gas (GHG) emissions.

2. *Charging optimization*

Scenario One (full BEB solution) and Scenario Three (mixed green fleet solution) will require the same number of on-route charging systems to supplement in-depot charging and energy transfer. Therefore, CUTRIC recommends pursuing Charging Strategy Two (depot with on-route) to ensure adequate energy transfer to power the fleet.

While Charging Strategy Two (depot with on-route charging) incurs additional infrastructure costs to achieve full decarbonization of the fleet, the land acquisition costs associated with procuring lands on which to install on-route charging systems are relatively insignificant (approximately half a million dollars, in 2023 dollars) compared to the overall cost of the project.

Additionally, the flexibility of additional energy and "fueling" for vehicles throughout the day minimizes the number of additional BEBs required to achieve a zero-carbon fleet. This smaller number of BEBs minimizes depot space requirements and fleet size. On-route charging also addresses the inflexibility of limited battery capacity, as the largest battery capacities available today still result in a significant portion of failed and unsuccessful blocks across the public transit agency service.

3. *Rollout and implementation plan*

To achieve the outcomes associated with any of the scenarios outlined above, including the preferred Scenario Three (mixed green fleet solution) by the aggressive timeline of 2040 for full fleet decarbonization, Brampton Transit will need to begin procuring, commissioning and deploying BEBs and/or FCEBs by the dozens on a per year basis beginning in 2026 across all three scenarios (full BEB, full FCEB, or mixed green fleet solution).

To implement Scenario Three (mixed green fleet solution), considering BEBs equipped with DFAH, Brampton Transit will need to procure 86 BEBs in 2026, ramping up to approximately 61 BEBs and 66 FCEBs in 2034 and by 2030, 10 BEBs and 20 FCEBs.

Figure 4 shows a summary of the conclusions of this report.

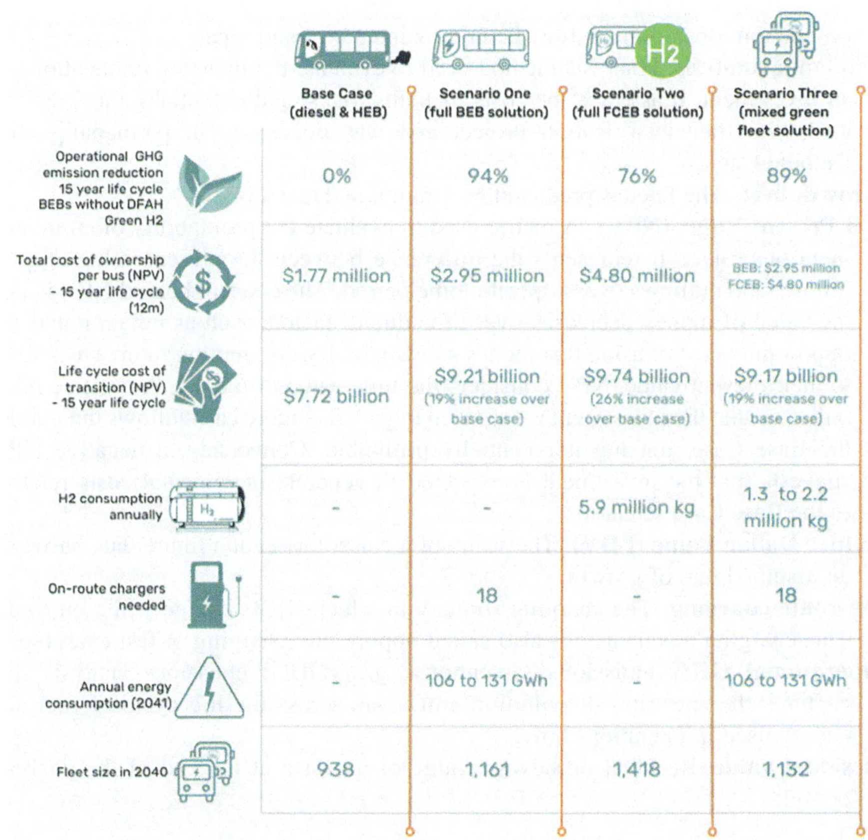

Fig. 4 Conclusions summary of scenarios

Glossary

Blocks A series of trips that are connected and assigned to a single vehicle.

Cash flow The movement of money into and out of a business' accounts over a specific period of time. It represents the cash generated or consumed by a company's operations and investments.

Current fleet Brampton Transit's fleet at the time of the study comprises 335 diesel buses that are 12 m long, 43 HEBs that are 12 m long, 90 HEBs that are 18 m long and eight BEBs that are 12 m long. The total fleet size of 476 buses is used in this analysis.

Depot charging The act of BEBs replenishing their batteries with electrical energy supplied by the transit facility, i.e., its garage. All BEBs are assumed to use depot charging to some extent.

Depot-only charging The charging strategy in which BEBs only recharge their batteries when inside their garage.

Environmental Life Cycle Analysis (LCA) Emission Life Cycle Analysis of all assets and fuel/energy used in operations $GHG^{+\ PLUS}$ module. It includes fuel production, operations, infrastructure, maintenance and repair.

Economic analysis Analysis method used to evaluate the financial implications of an investment. It assesses the costs, benefits, risks, and potential returns associated with the electrification project and was developed in partnership with Deloitte Canada.

Growth fleet The fleet as predicted by Brampton Transit in 2041.

Net Present Value (NPV) A metric used to evaluate the profitability of an investment or project. It represents the difference between the present value of cash inflows and outflows over a specific time period. The concept behind NPV is that the value of money decreases over time due to factors such as inflation and the opportunity cost of using that money elsewhere. By discounting future cash flows to their present value, NPV considers the time value of money. A positive NPV indicates that the investment is expected to generate more cash inflows than under the Base Case, making it potentially profitable. Conversely, a negative NPV suggests that the investment is expected to generate incremental costs relative to the Base Case scenario.

Future Dollar Value (FDV) The value of a current asset at a future date based on an assumed rate of growth.

On-route charging The charging strategy in which BEBs charge while on-route. This charging mechanism is also called opportunity charging or fast charging.

Operational GHG emissions Greenhouse gas (GHG) emissions analysis that captures the spectrum of polluting emissions across the life cycle of fuel and energy used in operations only.

Residual value Residual or salvage value of an asset at the end of the analysis period.

Scenario One (full BEB solution) In this scenario, the study assumes only battery electrified buses are used to achieve decarbonization. This scenario analyzes two charging strategies.

Scenario Two (full FCEB solution) In this scenario, the study assumes only hydrogen-fueled electrified buses (FCEBs) are used to achieve decarbonization.

Scenario Three (mixed green fleet solution) In this scenario, the study aims for an optimized combination of BEBs and FCEBs, reducing as much as possible the number of on-route chargers and charging episodes for BEBs, and the total number of buses for full-fleet decarbonization.

Spare ratio The ratio between the total fleet size and the number of peak vehicles on a typical weekday.Spare ratio $= \frac{\text{Total number of buses in the fleet}}{\text{Peak weekday number of buses}} = \frac{476}{331} = 1.44$

Total Cost of Ownership (TCO) The cost of purchasing and operating the vehicle and associated charging/refueling infrastructure over its entire lifespan, excluding labor costs.

References

1. The Paris Agreement (2015) The United Nations Framework Convention on Climate Change
2. City of Brampton (2021) Corporate greenhouse gas emissions. September
3. Natural Resources Canada (2010) 2008 Canadian vehicle survey update report. Natural Resources Canada, 6 October. Available: https://oee.nrcan.gc.ca/publications/statistics/cvs08/chapter2.cfm?attr=0
4. International Energy Agency (2021) Fuel economy in Canada, December 13
5. Prince Edward Island (2020) A climate change action plan for Prince Edward Island 2018–2023, 20 May. Available: https://www.princeedwardisland.ca/sites/default/files/publications/climatechange2018_f8.pdf
6. City of Brampton (2020) “Brampton’s Community Energy and Emissions Reduction Plan (CEERP). Available: https://pub-brampton.escribemeetings.com/filestream.ashx?DocumentId=258. Accessed on: 21 Aug 2023)
7. Colgan G. (2022) Roam transit, Parks Canada sign five-year $12.9 million deal, new electric buses to be purchased, 22 March. Available: https://www.rmotoday.com/banff/roam-transit-parks-canada-sign-five-year-deal-new-electric-buses-to-be-purchased-5186756
8. Government of Canada (2020) Ottawa to purchase new electric buses, 21 August 2023. Available: https://www.canada.ca/en/office-infrastructure/news/2020/12/ottawa-to-purchase-new-electric-buses.html
9. CUTRIC (2022) Canadian ZEB DatabaseTM: Canada’s zero-emission bus landscape and electrification readiness. Knowledge series. Quarterly Report #1
10. Union Internationale des Transports Publics (UITP) (2019) Policy brief, the impact of electric buses on urban life. Union Internationale des Transports Publics (UITP), June. Available: https://www.uitp.org/sites/default/files/cck-focus-papers-files/UITP-policybrief-June2019-V6%20WEB%20-%20OK_0.pdf
11. American Cancer Society (2015) Diesel exhaust and cancer risk, 21 August 2023. Available: https://www.cancer.org/healthy/cancer-causes/chemicals/diesel-exhaust-and-cancer.html
12. Sibanda E, Makaza N (2019) Health effects of diesel engine exhaust emissions exposure (DEEE) can mimic allergic asthma and rhinitis. Allergy, Asthma Clin Immunol 15(31):31
13. Laizāns A, Graurs I, Rubenis A, Utehin G (2016) Economic viability of electric public busses: regional perspective. Procedia Eng 134:316–321, January 1

14. MacMillan A, Turrentine J (2021) Global warming 101, 21 August 2023. Available: https://www.nrdc.org/stories/global-warming-101#:~:text=Though%20natural%20cycles%20and%20fluctuations,gas%2C%20which%20results%20in%20the
15. Xu Y, Ramanathan V, Victor DG (2018) Global warming will happen faster than we think. Nature 564(7734):30–32. December
16. New Flyer. Privacy policy. Available: https://www.newflyer.com/customer-service/#privacy-policy
17. New Flyer. Our next generation, zero-emission, fuel cell-electric bus. Available: https://www.newflyer.com/bus/xcelsior-charge-fc/
18. Petrunic J, Abotalebi E, Raj A (2020) Best practices and key considerations for transit electrification and charging infrastructure deployment to deliver predictable, reliable, and cost-effective Fleet systems. CUTRIC, Toronto
19. Howarth RW, Jacobson MZ (2021) How green is blue hydrogen? Energy Sci Eng 9:1676–1687, 12 August
20. Goran D (2016) The Elektromote, the world's first trolleybus, took to the streets of Berlin in 1882, 21 August 2023. Available: https://www.thevintagenews.com/2016/05/27/the-elektromote-the-worlds-first-trolleybus-took-to-the-streets-of-berlin-in-1882/
21. Bow J. The history of Toronto's trolley buses (1922–1993). Available: https://transittoronto.ca/trolleybus/9005.shtml. Accessed on: 26 Sept 2020
22. Poon L. Why U.S. cities aren't using more electric buses. Available: https://www.bloomberg.com/news/articles/2019-06-27/why-china-is-winning-the-electric-bus-race. Accessed on: 21 Aug 2023
23. Zhu S (2023) Will changing policies Hamper China's new energy bus and trucks markets in 2023? Available: https://interactanalysis.com/insight/will-changing-policies-hamper-chinas-new-energy-bus-and-trucks-markets-in-2023/
24. Nightingale A. Forget Tesla, it's China's e-buses that are denting oil demand. Available: https://www.bloomberg.com/news/articles/2019-03-19/forget-tesla-it-s-china-s-e-buses-that-are-denting-oil-demand. Accessed on: 21 Aug 2023
25. Prosser M. China's electric buses save more diesel than all electric cars combined. Available: https://singularityhub.com/2019/04/22/chinas-electric-buses-save-more-diesel-than-all-electric-cars-combined/. Accessed on: 21 Aug 2023
26. Feng X, Li CY, Chen DX, Tang J (2017) A method for defensing against multi-source Sybil attacks in VANET. Peer Peer Netw Appl 10(2):305–314
27. ZeEUS (2017) ZeEUS eBus Report-2 An updated overview of electric buses in Europe, 1 December 2019. Available: http://zeeus.eu/uploads/publications/documents/zeeus-ebus-report-2.pdf
28. Dolman M (2017) Fuel cell buses in Europe: latest developments and commercialisation pathway. Presented at the fuel cell buses conference, United Kingdom. Available: http://www.cte.tv/wp-content/uploads/2017/11/2-Madden_PDF.pdf
29. Stangarone T (2020) South Korean efforts to transition to a hydrogen economy. Clean Technol Environ Policy 23:1–8, 11 September
30. European Automobile Manufacturers' Association (2023) Vehicles in use in Europe 2023. , Available: https://www.acea.auto/files/ACEA-report-veshicles-in-use-europe-2023.pdf#page=16
31. Toronto Hydro (2022) Business rates and charges, 21 October. Available: https://www.torontohydro.com/for-business/rates
32. Gallucci M (2019) An increasingly urbanized Latin America turns to electric buses, 21 August 2023. Available: https://e360.yale.edu/features/an-increasingly-urbanized-latin-america-turns-to-electric-buses
33. Randall C (2020). Santiago de Chile receives 150 new BYD buses, 21 August 2023. Available: https://www.electrive.com/2020/06/30/santiago-de-chile-receives-150-new-byd-buses/
34. O'Boyle B (2022) Chile's first electric bus factory aims to ease fossil fuel dependency. Reuters, 22 July

35. Independent Electricity System Operator (2022) Global adjustment components and costs, 21 October. Available: https://www.ieso.ca/sector-participants/settlements/global-adjustment-components-and-costs
36. Tigue K (2019) U.S. electric bus demand outpaces production as cities add to their fleets, 21 August 2023. Available: https://insideclimatenews.org/news/14112019/electric-bus-cost-savings-health-fuel-charging
37. London Transit Commission (2020) Staff Report #4 electric bus assessment, 29 January, 3 March 2020. Available: http://www.londontransit.ca/staff-report-4-electric-bus-assessment-2/
38. Chard R, Hynes M, Lee B, Schnader J (2023) Zeroing in on ZEBs. Available: https://calstart.org/wp-content/uploads/2023/02/Zeroing-in-on-ZEBs-February-2023_Final.pdf
39. CBC News (2019) B.C. Transit switching entire fleet to electric buses, 29 July, 3 December 2019. Available: https://www.cbc.ca/news/canada/british-columbia/bc-transit-electric-bus-fleet-1.5229665
40. Fuel Cells Works (2018) Hyundai puts hydrogen powered buses into service in Seoul, 21 August 2023. Available: https://fuelcellsworks.com/news/hyundai-puts-hydrogen-powered-buses-into-service-in-seoul/
41. Hampel C (2022) S. Korea to acquire 624 H2 fuel cell buses by 2025. Rabbit Publishing, Berlin
42. ManThey N (2022) 1,000 hydrogen buses in a bid of Busan to become next eco-city. Rabbit Publishing, Berlin
43. Hydrogen Council (2017) hydrogen, scaling up. Toyota Motor Corporation, Brussels
44. Harding R (2021) High costs dog Tokyo's hydrogen buses. *Financial Times*, July
45. Johnson C, Nobler E, Eudy L, Jeffers M (2020) Financial analysis of battery electric transit buses. National Renewable Energy Laboratory. Available: https://afdc.energy.gov/files/u/publication/financial_analysis_be_transit_buses.pdf
46. California Air Resources Board (2018) California transitioning to all-electric public bus fleet by 2040, 21 August 2023. Available: https://ww2.arb.ca.gov/news/california-transitioning-all-electric-public-bus-fleet-2040
47. Red River College (2017) Zero emission – electric transit bus in Manitoba. In: Prototype electric transit bus development and demonstration – final report, June. Available: https://www.rrc.ca/vtec/wp-content/uploads/sites/78/2019/05/Prototype-Electric-Bus-Development-and-Demonstration-Final-report-1pxb35m.pdf. Accessed on: 21 Aug 2023
48. MARCON (2016) Electric bus feasibility study for the city of Edmonton, June. Available: https://www.edmonton.ca/documents/transit/ETS_Electric_Feasibility_Study.pdf. Accessed on: 21 Aug 2023
49. CUTRIC-CRITUC (2021) Battery electric buses: the promise versus realities of zero-emissions transit technologies, 29 March. In: 1st annual ZEB Technologies & Transit Operations conference
50. Chan K (2019) TransLink has received its first ever electric-battery bus, 3 December 2019. Available: https://dailyhive.com/vancouver/translink-receives-first-electric-battery-bus
51. Climate Change Accountability Act (ed) (2007) Government of British Columbia, Canada
52. Yakub M (2023) Ottawa green-lights purchase of 350 electric transit buses. Electric Autonomy Canada, 6 February
53. City of Toronto (2016) TTC ridership – all day weekday for surface routes. C. o. Toronto, Toronto
54. Sokic N (2023) TTC orders up to 1,162 Nova bus, new flyer electric buses. Sustainable Biz Canada
55. City of Edmonton (2018) Edmonton's electric vehicle strategy, 1 December 2019. Available: https://www.edmonton.ca/city_government/documents/PDF/EdmontonElectricVehicleStrategy.pdf

56. Société de transport de Montréal (2015) City mobility—the first fully electric bus made in Québec unveiled, 24 November, 26 April 2021. Available: http://www.stm.info/en/press/news/2015/city-mobility%2D%2D-the-first-fully-electric-bus-made-in-quebec-unveiled
57. Carty M (2020) $177M to be spent to electrify Guelph transit buses, build bus storage facility. Available: https://globalnews.ca/news/6422264/electric-guelph-transit-buses-funding/. Accessed: 29 Sept 2020
58. Mazur A (2019). Kingston receives over $500K in federal funding for new electric buses, 21 August 2023. Available: https://globalnews.ca/news/5854148/kingston-500k-new-electric-buses/
59. LeBel J (2020) Study completion marks major step in London Transit Commission push for zero-emissions bus fleet, 21 August 2023. Available: https://globalnews.ca/news/7484149/london-transit-zero-emissions-bus-plan/
60. MacDonald D (2020) Greater Sudbury's long-term plan will see an all-electric fleet of vehicles, including city buses, 21 August 2023. Available: https://northernontario.ctvnews.ca/greater-sudbury-s-long-term-plan-will-see-an-all-electric-fleet-of-vehicles-including-city-buses-1.5225917
61. Cooke E (2021) Transition to zero-emission technical evaluation report. Winnipeg Transit. Accessed on: 25 Mar 2021. Available: https://winnipegtransit.com/assets/2788/Transition_to_Zero_Emission_Technology_Report_-_Rev1.pdf
62. Eudy L, Post M (2014) BC transit fuel cell bus project: evaluation results report. National Renewable Energy Laboratory, Golden. Available: https://www.nrel.gov/docs/fy14osti/60603.pdf. Accessed on: 21 Aug 2023
63. MiWay (2020) MiWay 2020–2023 business plan & 2020 budget. MiWay, Mississauga. . Available: https://web.mississauga.ca/wp-content/uploads/2019/11/08151657/D-MiWay.pdf. Accessed on: 21 Aug 2023
64. CUTRIC (2022) Toronto Transit Commission (TTC) fleet electrification: feasibility analysis & simulation-based planning. Canadian Urban Transit Research and Innovation Consortium (CUTRIC), Toronto, 31 December
65. City of Edmonton (2023) Hydrogen and electric buses. Available: https://www.edmonton.ca/projects_plans/transit/electric-buses
66. Millick F (2023) Halifax seeks study on adding hydrogen fuel cell buses to transit fleet, 27 January. Available: https://www.cbc.ca/news/canada/nova-scotia/halifax-transit-hydrogen-fuel-cell-buses-1.6727199. Accessed on: 12 Feb 2024
67. Llamas S (2023) AC transit zero-emission bus transition interview. CUTRIC, Toronto
68. Olsen N, Kliewer N, Wolbeck L (2020) A study on flow decomposition methods for scheduling of electric buses in public transport based on aggregated time–space network models. Cent Eur J Oper Res 30:883–919. https://link.springer.com/article/10.1007/s10100-020-00705-6#citeas
69. López-Ibarra JA, Gaztañaga H, Saez-de-Ibarra A, Camblong H (2020) Plug-in hybrid electric buses total cost of ownership optimization at fleet level based on battery aging. Appl Energy 280:115887
70. Yusof NK, Abas PE, Mahlia TMI, Hannan MA (2021) Techno-economic analysis and environmental impact of electric buses. World Electr Vehicle J 12(1):31
71. Klose C et al (2020) All-hydrocarbon MEA for PEM water electrolysis combining low hydrogen crossover and high efficiency. Adv Energy Mater 10(14):1903995
72. Abdelaty H, Mohamed M (2021) A prediction model for battery electric bus energy consumption in transit. Energies 14(10):2824
73. The Thomas D. Larson Pennsylvania Transportation Institute and Bus Testing and Research Center (2015) Federal transit bus test – New Flyer XE40. The Thomas D. Larson Pennsylvania Transportation Institute, University Park, PALTI-BT-R1405. Available: http://apps.altoonabustest.psu.edu/buses/reports/458.pdf?1441118410. Accessed on: 21 Aug 2023
74. The Thomas D. Larson Pennsylvania Transportation Institute and Bus Testing and Research Center (2017) Federal transit bus test – Proterra, Inc. Catalyst E2. The Thomas D. Larson

Pennsylvania Transportation Institute, University Park, PALTI-BT-R1706-P. Available: http://apps.altoonabustest.psu.edu/buses/reports/480.pdf?1521553696. Accessed on: 21 Aug 2023
75. The Thomas D. Larson Pennsylvania Transportation Institute and Bus Testing and Research Center (2018). FEDERAL TRANSIT BUS TEST – New Flyer XHE40. The Thomas D. Larson Pennsylvania Transportation Institute and Bus Testing and Research Center, University Park. Available: http://apps.altoonabustest.psu.edu/buses/reports/501.pdf?1547230960
76. Eudy L, Post M, Jeffers M (2016) Zero emission bay area (ZEBA) fuel cell bus demonstration results: fifth report. National Renewable Energy Laboratory, Golden. Available: https://www.nrel.gov/docs/fy16osti/66039.pdf
77. Burress T (2013) Benchmarking state-of-the-art technologies, 21 August 2023. Available: https://www.energy.gov/sites/prod/files/2014/03/f13/ape006_burress_2013_o.pdf
78. U.S. Department of Transportation (2023) Benefit-cost analysis guidance for discretionary grant programs, 5 December
79. University of Calgary (2022) Transit signal priority in connected and autonomous vehicle environments. University of Calgary, Calgary
80. Hanlin J, Reddaway D, Lane J (2018) Battery electric buses – state of the practice – a synthesis of transit practice. Center for Transportation and the Environment, Atlanta. [Online]. Available: https://www.plugincanada.ca/wp-content/uploads/2019/02/25061.pdf
81. Elrayies GM (2017) Aerial ropeways as catalysts for sustainable public transit in Egypt. Int J Innov Res Sci Eng Technol 6(8):15542–15555
82. Ferreras J (2019) In 2013, 6 were killed when an Ottawa double-decker bus hit a train. Distraction was one cause. Global News 12 January
83. SAE International (2023) Electric vehicle power transfer system using conductive automated connection devices infrastructure-mounted pantograph (cross-rail) connection: J3105/1_202305. SAE International, Warrendale
84. Transport Canada (2010) Government partnership celebrates unveiling of new ZÜM bus for Brampton, 20 April
85. Liu X, Reddi K, Elgowainy A, Lohse-Busch H, Wang M, Rustagi N (2020) Comparison of well-to-wheels energy use and emissions of a hydrogen fuel cell electric vehicle relative to a conventional gasoline-powered internal combustion engine vehicle. Int J Hydrog Energy 45(1):972–983
86. The EV Report (2023) InductEV partners with ChargePoint to expand wireless EV charging solutions, 6 April
87. Momentum Dynamics (2020 Link Transit achieves 50 MWh of energy delivered wirelessly to electric bus fleet, 21 August 2023. Available: https://www.masstransitmag.com/bus/vehicles/hybrid-hydrogen-electric-vehicles/press-release/21139036/momentum-dynamics-link-transit-achieves-50-mwh-of-energy-delivered-wirelessly-to-electric-bus-fleet
88. Masquelier M (2017) Wireless charging infrastructure for battery electric buses in public transportation. In: APTA sustainability conference, Minneapolis, MN, USA
89. Wang X et al (2021) A review of modeling, acquisition, and application of lithium-ion battery impedance for onboard battery management. eTransportation 7. https://doi.org/10.1016/j.etran.2020.100093
90. Morris DZ (2017) Driverless cars will be part of a $7 trillion market by 2050, 14 January. Available: https://fortune.com/2017/06/03/autonomous-vehicles-market/
91. Canadian Urban Transit Research & Innovation Consortium (2020) Pan-Canadian electric bus demonstration & integration trial: phase I & II, 15 March. Available: https://cutric-crituc.org/project/pan-canadian-electric-bus-demonstration-integration-trial/
92. Eudy L (2017) U.S. fuel cell projects. Available: https://www.nrel.gov/hydrogen/fuel-cell-bus-evaluation.html
93. Blanco S (2019) Proterra ready for electric bus battery leasing with $200-million credit facility, 21 August 2023. Available: https://www.forbes.com/sites/sebastianblanco/2019/04/18/proterra-ready-for-electric-bus-battery-leasing-with-200-million-credit-facility/#2088e8432314

94. Field K (2018) BYD introduces new $200 million electric bus leasing program in partnership with generate capital, 21 August 2023. Available: https://cleantechnica.com/2018/07/10/byd-introduces-new-200-million-electric-bus-leasing-program-in-partnership-with-generate-capital/
95. Eudy L, Matthew P (2018) Fuel cell buses in U.S. transit fleets: current status 2018. National Renewable Energy Laboratory. Available: https://www.nrel.gov/docs/fy19osti/72208.pdf
96. Eudy L, Post M (2018) Fuel cell buses in US transit fleets: current status 2018, December, 21 August 2023. Available: https://www.nrel.gov/docs/fy19osti/72208.pdf
97. Mathison S (2020) How advanced hydrogen fueling protocols can improve fueling performance & H2 station design. Available: https://www.nrel.gov/docs/fy20osti/77368.pdf
98. Alternative Fuels Data Center (2019) SunLine transit demonstrates excellence in zero emission transit, 26 November 2019. U.S. Department of Energy. Available: https://afdc.energy.gov/case/3086
99. Kuo S (2021) Metro's battery bus plan hinges on need to grow service and revenue," The Urbanist, January
100. Alternative Fuels Data Center (2020) Hydrogen production and distribution, 21 August 2023. Available: https://afdc.energy.gov/fuels/hydrogen_production.html
101. IEA Bioenergy (2018) Hydrogen from biomass gasification, December. Available: https://www.ieabioenergy.com/wp-content/uploads/2019/01/Wasserstoffstudie_IEA-final.pdf
102. The Fuel Cells and Hydrogen Joint Undertaking (2012) Urban buses: alternative powertrains for Europe. The Fuel Cells and Hydrogen Joint Undertaking (FCH-JU). Available: https://www.gppq.fct.pt/h2020/_docs/brochuras/fch-ju/20121029%20urban%20buses,%20alternative%20powertrains%20for%20europe%20-%20final%20report_0.pdf. Accessed on: 18 Nov 2019
103. Ewing M, Israel B, Jutt T, Talebian H, Stepanik L (2020) Hydrogen on the path to net-zero emissions, costs and climate benefits. Pembina Institute. Available: https://www.pembina.org/reports/hydrogen-climate-primer-2020.pdf. Accessed on: 21 Aug 2023
104. Natural Resources Canada (2014) Learn the facts: fuel consumption and CO2, 4 May. Available: https://www.nrcan.gc.ca/sites/www.nrcan.gc.ca/files/oee/pdf/transportation/fuel-efficient-technologies/autosmart_factsheet_6_e.pdf
105. U.S. Department of Energy. Liquid hydrogen delivery. Available: https://www.energy.gov/eere/fuelcells/liquid-hydrogen-delivery
106. Air Products (2019) Air products company info, 3 December. Available: http://www.airproducts.ca/
107. Sarnia-Lambton Economic Partnership (2024) Ontario's hydrogen hub
108. Burke A, Sinha AK (2020) Technology, sustainability, and marketing of battery electric and hydrogen fuel cell medium-duty and heavy-duty trucks and buses in 2020–2040, NCST-UCD-RR-20-09. University of California, Davis
109. Ballard Which zero-emission bus type is best for your fleet? ed. https://www.ballard.com/docs/default-source/motive-modulesdocuments/beb-or-fceb_final_feb_22_2019.pdf?sfvrsn=892c080_2#:~:text=BEB%20FCEB%20WHICH%20ZERO%2DEMISSION,IS%20BEST%20FOR%20YOUR%20FLEET%3F&text=BEBS%20excel%20for%20agencies%20with%20short%20routes%20and%20moderate%20schedules.&text=BEBS%20excel%20where%20depots%20have%20few%20buses%20or%20ample%20parking%20space.&text=FCEBS%20excel%20where%20depots%20have%20many%20buses%20or%20are%20space%20constrained
110. Stephens B (2022) Canadian Pacific's hydrogen-powered locomotive makes first revenue run, 21 August 2023. Available: https://www.trains.com/trn/news-reviews/news-wire/canadian-pacifics-hydrogen-powered-locomotive-makes-first-revenue-run/
111. Alstom (2023) Alstom concludes the successful demonstration of the first commercial service hydrogen-powered train in North America. Available: https://www.alstom.com/press-releases-news/2023/10/alstom-concludes-successful-demonstration-first-commercial-service-hydrogen-powered-train-north-america

112. Alstom (2023) Alstom's Coradia iLint, the world's first hydrogen-powered passenger train, will demonstrate green traction in Quebec. Available: https://www.alstom.com/press-releases-news/2023/2/alstoms-coradia-ilint-worlds-first-hydrogen-powered-passenger-train-will-demonstrate-green-traction-quebec
113. King County Metro (2022) King County Metro transit's zero emission fleet transition plan, May
114. Alonso-Mora J, Samaranayake S, Wallar A, Frazzoli E, Rus D (2017) On-demand high-capacity ride-sharing via dynamic trip-vehicle assignment. Proc Natl Acad Sci USA 114(3): 462–467, 17 January
115. Eudy L, Jeffers M Zero-emission bus evaluation results: King County Metro battery electric buses. National Renewable Energy Laboratory. Available: https://www.transit.dot.gov/sites/fta.dot.gov/files/docs/research-innovation/115086/zero-emission-bus-evaluation-results-king-county-metro-battery-electric-buses-fta-report-no-0118.pdf. Accessed on: 19 May 2021
116. Quarles KKN, Mohamed M (2020) Costs and benefits of electrifying and automating bus transit fleets. Sustainability 12:3977
117. BestMile (2017) BestMile, CarPostal & Navya to present very first live autonomous shuttle demonsration at cebit. BestMile, Lausanne
118. Baker DR (2023) Will white, blue and green hydrogen fuel a clean-energy future, or fizzle? Available: https://www.bloomberg.com/news/articles/2023-10-11/hydrogen-fuel-of-the-future-or-fizzling-climate-change-solution
119. Bank of Canada (2024) Historical noon and closing rates, 22 January. Available: https://www.bankofcanada.ca/rates/exchange/legacy-noon-and-closing-rates/
120. Bank of Canada (2024) Monthly exchange rates, 22 January. Available: https://www.bankofcanada.ca/rates/exchange/monthly-exchange-rates/#download
121. CUTRIC (2023) Pan-Canadian hydrogen fuel cell electric bus demonstration and integration trial, 21 August. Available: https://cutric-crituc.org/marquee-projects/pan-canadian-hydrogen-fuel-cell-electric-bus-demonstration-and-integration-trial/
122. Government of Ontario (2023) Ontario's low-carbon hydrogen strategy, January 25
123. IESO (2022) Connecting today. Powering tomorrow, 14 March. Available: https://www.ieso.ca/
124. Ministry of Energy (2022) Accelerating growth in hydrogen energy through electricity rate options, 14 April
125. Government of Ontario (2023) Ontario boosting its electricity grid with hydrogen. Available: https://news.ontario.ca/en/release/1003866/ontario-boosting-its-electricity-grid-with-hydrogen
126. London Transit Commission (2021) 2022–2031 capital budget program. Available: https://www.londontransit.ca/staff-report-5-2022-2031-capital-budget-program-2/
127. Chan K (2019) TransLink installs its first electric-battery bus charger outside SkyTrain station. Urbanized [Transportation]. Available: https://dailyhive.com/vancouver/translink-battery-electric-bus-charger-22nd-street-station-april-2019
128. Chan K (2023) Big growing pains for TransLink to electrify buses and expand services, 5 July
129. Wanek-Libman M (2023) STM receives first electric bus. Available: https://www.masstransitmag.com/bus/vehicles/hybrid-hydrogen-electric-vehicles/article/21114528/stm-receives-first-electric-bus
130. Société de transport de Montréal (2023) Bus network electrification. Available: https://www.stm.info/en/about/major_projects/major-bus-projects/bus-network-electrification
131. City of Kingston (2022) Electrify city's fleet, 21 August 2023. Available: https://dev2.cityofkingston.ca/apps/councilpriorities/climate_electrifyfleet.html
132. City of Kingston (2019) Municipalities for Climate Innovation Program funding supports city's strategic priority of incrementally electrifying vehicle fleet. Available: https://www.cityofkingston.ca/-/municipalities-for-climate-innovation-program-funding-supports-city-s-strategic-priority-of-incrementally-electrifying-vehicle-fleet

Sustainable Electrocatalyst for PEM Water Electrolyzers

Mehmet Fatih Kaya, Murat Kıstı, Bulut Hüner, Emre Özdoğan, Marise Conağası, Muhammed Ali Durmaz, Eda Nur Çakıraslan, Tayyar Eşiyok, Yakup Ogün Süzen, Akif Taşkın, Süleyman Uysal, and Nesrin Demir

1 Introduction

Hydrogen is considered one of the most efficient options for energy transportation, as it can be obtained from various energy sources [1]. Hydrogen is produced by four main methods: hydrocarbon reforming, biomass processing, hydrocarbon pyrolysis, and water electrolysis. Electrochemical water electrolysis stands out among various hydrogen production methods, demonstrating the capacity to achieve high purity levels (up to 99.999%) of hydrogen. In addition, water electrolysis provides a major

M. F. Kaya (✉) · S. Uysal
Engineering Faculty, Energy Systems Engineering Department, Heat Engineering Division, Erciyes University, Kayseri, Turkey

Erciyes University H2FC Hydrogen Energy Research Group, Kayseri, Turkey

ArGePark Research Building, Erciyes University, Kayseri, Turkey

BATARYASAN Enerji San. ve Tic. A. Ş., Erciyes Teknopark, Kayseri, Turkey
e-mail: kayamehmetfatih@erciyes.edu.tr

M. Kıstı · M. A. Durmaz
Engineering Faculty, Energy Systems Engineering Department, Heat Engineering Division, Erciyes University, Kayseri, Turkey

Erciyes University H2FC Hydrogen Energy Research Group, Kayseri, Turkey

ArGePark Research Building, Erciyes University, Kayseri, Turkey

Graduate School of Natural and Applied Sciences, Energy Systems Engineering Department, Erciyes University, Kayseri, Turkey

B. Hüner
Erciyes University H2FC Hydrogen Energy Research Group, Kayseri, Turkey

ArGePark Research Building, Erciyes University, Kayseri, Turkey

Directorate of Research and Innovation, Osmaniye Korkut Ata University, Osmaniye, Turkey

A. Kovač (ed.), *Challenges and Solutions in the Hydrogen Value Chain*,
https://doi.org/10.1007/978-3-031-69051-8_3

contribution to the production of green hydrogen in combination with alternative sources of energy. There are currently four water electrolysis technologies available: Polymer Electrolyte Membrane Water Electrolyzer (PEMWE), alkaline water electrolysis (AWE), anion exchange membrane water electrolysis (AEMWE), and solid oxide water electrolysis (SOWE) [2]. Among these electrolyzer technologies, PEMWE stands out as a prominent technology in terms of sustainability. PEMWE is a bright future technology for the sustainable generation of hydrogen, successfully integrating electricity obtained from renewable or alternative energy sources like wind and solar [3]. PEMWE, which uses a thin, solid sulfonated polystyrene

E. Özdoğan
Engineering Faculty, Energy Systems Engineering Department, Heat Engineering Division, Erciyes University, Kayseri, Turkey

Erciyes University H2FC Hydrogen Energy Research Group, Kayseri, Turkey

ArGePark Research Building, Erciyes University, Kayseri, Turkey

BATARYASAN Enerji San. ve Tic. A. Ş., Erciyes Teknopark, Kayseri, Turkey

Graduate School of Natural and Applied Sciences, Energy Systems Engineering Department, Erciyes University, Kayseri, Turkey

Engineering Faculty, Electrical-Electronics Engineering Department, Erciyes University, Kayseri, Turkey

M. Conağası · T. Eşiyok
Engineering Faculty, Energy Systems Engineering Department, Heat Engineering Division, Erciyes University, Kayseri, Turkey

Erciyes University H2FC Hydrogen Energy Research Group, Kayseri, Turkey

ArGePark Research Building, Erciyes University, Kayseri, Turkey

BATARYASAN Enerji San. ve Tic. A. Ş., Erciyes Teknopark, Kayseri, Turkey

Graduate School of Natural and Applied Sciences, Energy Systems Engineering Department, Erciyes University, Kayseri, Turkey

E. N. Çakıraslan
Erciyes University H2FC Hydrogen Energy Research Group, Kayseri, Turkey

ArGePark Research Building, Erciyes University, Kayseri, Turkey

Engineering Faculty, Electrical-Electronics Engineering Department, Erciyes University, Kayseri, Turkey

Y. O. Süzen · A. Taşkın
Erciyes University H2FC Hydrogen Energy Research Group, Kayseri, Turkey

ArGePark Research Building, Erciyes University, Kayseri, Turkey

N. Demir
Engineering Faculty, Energy Systems Engineering Department, Heat Engineering Division, Erciyes University, Kayseri, Turkey

Erciyes University H2FC Hydrogen Energy Research Group, Kayseri, Turkey

ArGePark Research Building, Erciyes University, Kayseri, Turkey
e-mail: nkayatas@erciyes.edu.tr

membrane as the electrolyte to transport protons [4]. This membrane provides advantages such as isolating the cathode and anode electrodes, resulting in high current density (>2 A.cm^{-2}), high conversion efficiency (80–90%), and high hydrogen purity (>99.99%). Moreover, PEMWE exhibits the capability to produce high-purity hydrogen at relatively lower temperature ranges, around 70–90 °C, compared to other electrolyzers [5]. PEMWE consists of a catalyst-coated membrane (CCM), anode and cathode electrodes, bipolar plates (BP), and gas diffusion layers (GDL). The CCM is produced by loading the catalyst onto the membrane [6]. GDL is employed to facilitate the diffusion of gases to the electrode surfaces. On the other hand, BP serves as a material that transfers gases and liquids into and out of the cell while also acting as a current collector. The electrolysis process of PEMWE comprises two basic electrochemical semi-reactions (at the anode: OER and at the cathode: HER) [7]. Theoretically, only a voltage of 1.23 V is required for PEMWE to achieve hydrogen production. However, the electrolysis process typically requires a much larger voltage due to overpotentials at both the cathode and anode (generally around 1.8–2.0 V). Therefore, different electrocatalysts are employed to mitigate the overpotentials for both HER and OER [4]. Electrocatalysts have an important role in controlling the reaction rate and efficiency of the electrolysis process, which affects the efficiency of the overall system. The most fundamental factor to consider in order to maximize the hydrogen production performance of PEMWE is electrocatalyst selection [8]. Electrocatalysts are typically selected for specific characteristics such as pore structure, particle size, high electrical conductivity, high active surface area, and corrosion stability under oxidative conditions [9]. The currently available structures that can be utilized as anode catalysts for PEMWE are primarily based on Ir, Ru, and similar noble metallic catalysts. Pt is used as a cathode-side catalyst in PEMWEs and is considered an advanced and benchmark catalyst due to its exceptional activity and long-term stability [10]. However, catalyst costs hinder the commercialization and industrialization of PEMWE technology. To achieve a balance between cost, activity, and stability in electrocatalyst design, the following strategies have emerged: (a) addition of non-noble metal components to decrease the noble metal content in the electrocatalyst, (b) development of support catalysts with high surface area and good conductivity to improve Ir mass activity, (c) regulation of the local environment and electronic structure of the electrocatalyst to enhance its intrinsic activity. Moreover, a limited number of studies have been conducted using high-performance and low-cost HER electrocatalysts in PEMWE instead of precious metal catalysts. There is a great deal of interest in developing alternative catalysts using elements abundant in the Earth's crust to replace Pt. Transition metal elements, particularly Fe, Ni, Co, Mo, Cu, and W, are often used for Pt-free HER electrocatalysts. These catalysts are considered as potential replacements for Pt and have already been used to produce highly active electrocatalysts for HER [11]. However, Pt Group Metals (PGMs) remain on the updated EU 2020 list of Critical Raw Materials, underscoring the imperative for their substitution and recycling [12]. Although there has been a significant increase in annual Pt production, Ir is predominantly a byproduct of Pt mining, making its availability more limited. This limits the widespread use of Ir for large-scale PEMWE systems. Another important issue is that Pt is highly dependent on imports and 71% of Pt is supplied to the EU from South Africa. Moreover, only 21% of the demand for PGMs can be met from secondary raw materials [13]. To overcome the challenges determined by the use of critical feedstocks and secure access to sustainable catalysts for PEMWE, there is an ongoing and expanding research effort aimed at minimizing the use of PGMs in catalysts or replacing them entirely with cost-effective and abundantly available metals. In this section, we discuss the recent advances in sustainable electrocatalysts for PEMWEs, such as efficiency, durability, and their challenges. The key challenges in developing sustainable electrocatalysts for PEMWEs include maintaining high activity, improving stability, and reducing costs. Ongoing research and development aim to address these challenges and contribute to the widespread adoption of water electrolysis for hydrogen production. Therefore, by addressing challenges and highlighting recent advances, we present a review of the literature that presents a broad overview of the present state of sustainable electrocatalysts for PEMWEs and their potential impact on the future of hydrogen production.

2 Basics of PEM Water Electrolyzers

2.1 Principle of Operation

In PEMWE, the anode and cathode catalytic layers are separated from each other by a solid electrolyte. This solid electrolyte is known as Nafion membrane, which becomes strongly acidic when immersed in water. Catalytic layers can be formed by an electroless coating or hot pressing on both sides of the membrane [14]. In a PEMWE system, the anode is the electrode where water is oxidized, and oxygen gas is produced. The process involves the splitting of water molecules into protons (H^+) and oxygen molecules (O_2). At the anode, water (H_2O) is oxidized to produce protons (H^+), electrons (e^-), and oxygen gas (O_2). At the cathode side, protons and electrons produced at the anode side combine to form hydrogen gas (H_2). Protons are transferred to the cathode through the PEM, and electrons are transferred to the cathode by the exterior electrical connection to form hydrogen [1]. The current density applied by the external circuit to the cell is in the range of 10.000–20.000 A/m^2, which is approximately five times higher than AWE. PEMWE can operate in the temperature range of 50–80 °C and at pressures lower than 5 MPa [14]. Figure 1 illustrates the basic representation of PEMWE.

The anode, cathode and total reactions occurring in PEMWE are given in Eqs. (1), (2), and (3) as follows:

Anode Reaction: H_2O is oxidized at the anode and O_2, H^+ and e^- are formed.

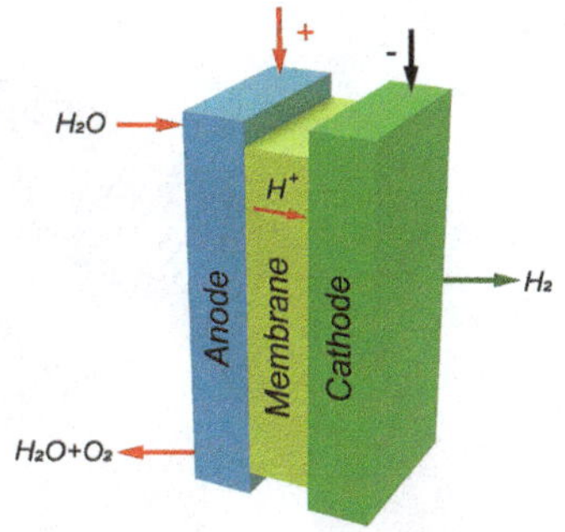

Fig. 1 Schematic view of a single-cell PEMWE

$$2H_2O \rightarrow O_2 + 4H^+ + 4e^- \quad (1)$$

Cathode Reaction: H^+ and e^- are reduced at the cathode and molecular hydrogen (H_2) is formed.

$$4H^+ + 4e^- \rightarrow 2H_2 \quad (2)$$

Total Reaction:

$$2H_2O \rightarrow 2H_2 + O_2 \quad (3)$$

The polymer electrolyte membrane in PEMWE separates the cathode and anode components, allowing only protons to pass through while blocking the electrons and oxygen. This ensures the separation of hydrogen and oxygen production [15].

2.2 *Components of a PEMWE*

The main components of a PEMWE include the membrane electrode assembly (MEA), GDL, BP, and compression plate (CP). However, there is an urgent need to research advanced materials suitable for PEMWE due to the high capital cost of PEMWE stacks. The different PEMWE systems may have additional components or variations in the arrangement of these basic elements. Figure 2 presents a simplified overview of the main components of a PEMWE. The actual design may vary depending on the specific manufacturer and application.

According to Fig. 2, the actual design and arrangement of these components may vary among manufacturers and applications. Moreover, advancements in technology may lead to variations or the inclusion of additional properties for good performance, durability, or stability of PEMWEs.

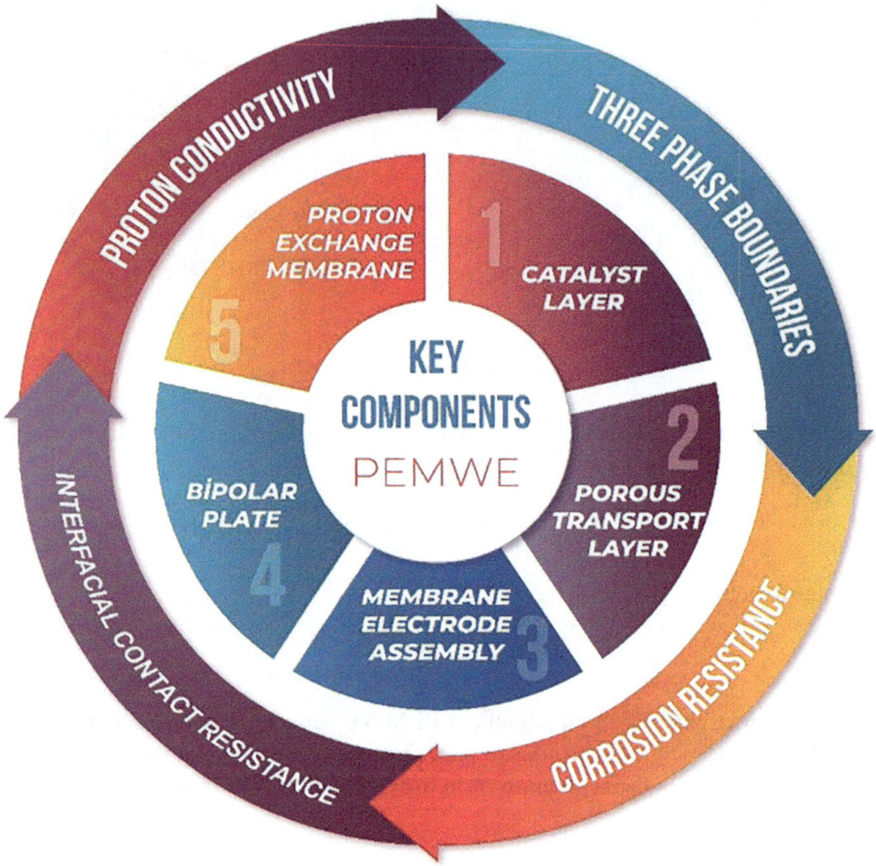

Fig. 2 Schematic illustration of key components in PEMWE

2.2.1 CP

The CP is typically made of aluminum alloy, which is used to secure the complete electrolysis cell. CP is located at the top of the cell and is responsible for compressing the cell stack to ensure that the BPs are in good contact with the MEA. CP, which is an important component of PEMWE, ensures tight packing of the cell stack and compression of the MEA to the correct thickness.

2.2.2 BP

BP is a multi-functional component in the PEMWE stacks. It contains flow channels that provide a constant water flow to the MEA surface. In addition, it rapidly provides gaseous product removal and mechanical support. Furthermore, BPs

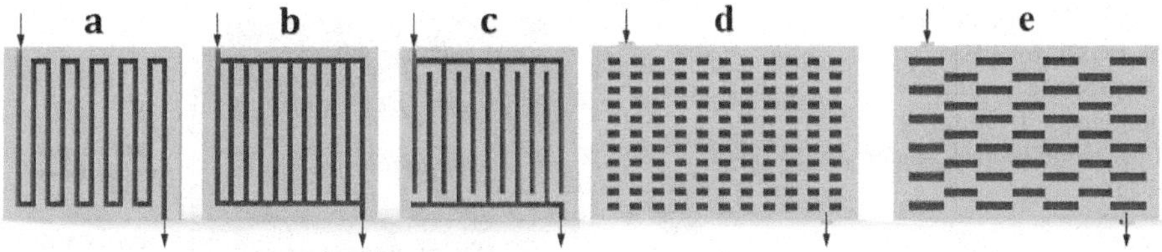

Fig. 3 Different BP geometry designs: (**a**) Serpentine (**b**) Parallel (**c**) Interdigitated (**d**) Pin (**e**) Cascade design. (Reproduced with permission from ref. 16. Copyright 2019, Elsevier [16])

serve the crucial function of isolating individual stacked electrolyzers while facilitating electrical contact and conduction of heat between stacked electrolyzers. The life of a cell stack depends on how the gases enter and exit the fuel cell. Therefore, the power output of the cell stack largely depends on the design of the BPs. Figure 3 shows different BP geometry designs used in PEMWEs.

Many of these configurations are categorized as serpentine, parallel, interlocking, pin, and cascade. However, among all other types of BP geometry, parallel design is the least complex [16, 17]. In the parallel design depicted in Fig. 3, the liquid flow is uniformly distributed along the flow channels with minimal pressure loss [18]. The manufacturing method employed by BP is a critical determinant of the cost, weight, and volume of the fuel cell. There are several methods for producing bipolar plates, including casting, machining, and stamping. Therefore, BPs are also being produced by the innovative manufacturing process of additive manufacturing [19].

2.2.3 GDL

An important component of PEMWE that enables efficient mass transfer of gases and liquids between electrodes and BP is GDL. Therefore, it is located between the MEA and the BPs, serving the essential function of uniform distribution of reactant gases across the catalyst layer's surface. The selection of materials for the GDL is crucial in terms of cost, performance, and fuel cell durability. GDL may consist of diverse materials, including carbon fiber paper, carbon cloth, metal mesh, porous metal foam, and nonwoven structures based on carbon fiber [6, 20]. Figure 4 gives an overview of GDL for PEMWE systems.

In the anode side of PEMWEs, various types of sintered porous titanium structures such as felts, meshes, and foams are commonly used as GDLs. Whereas meshes are an economical option for GDLs, their performance in electrolyzers is not as high as that of sintered structures or felts. Sintered titanium plates have limitations in terms of both area and thickness, but they are capable of achieving high performance [22]. Creating cost-effective GDLs with optimal pore size, porosity, surface properties, and thickness remains a challenge for PEMWEs.

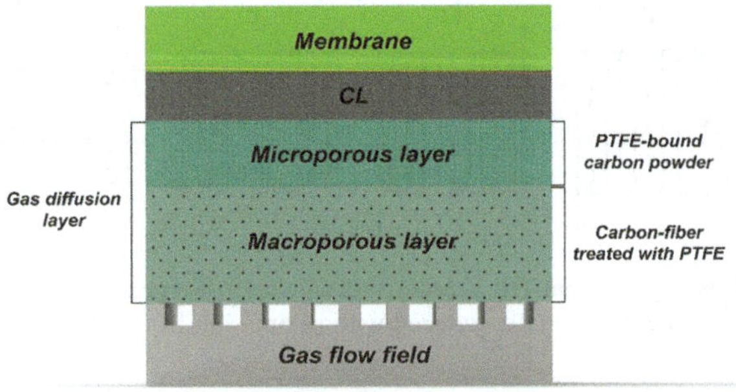

Fig. 4 Schematic diagram of a GDL for PEMWE. (Reproduced with permission from ref. 21. Copyright 2006, Elsevier [21])

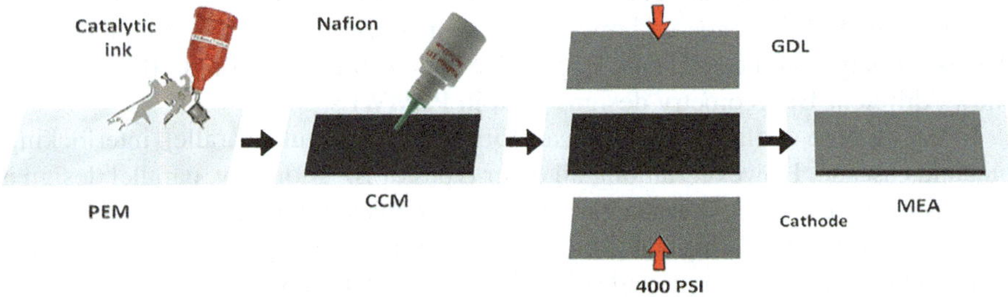

Fig. 5 Schematic illustration of MEA fabrication

2.2.4 MEA

MEA is located between two electrodes: the cathode catalyst layer (CCL) and the anode catalyst layer (ACL). The CCL and ACL consist of electrocatalysts and proton exchange ionomers. GDLs are fabricated on both the cathode and anode sides of the MEA, and BPs are attached to the outside of the GDLs [23]. The MEA is responsible for the electrochemical reactions that occur in fuel cells. The MEA is a complex system that requires careful design and optimization. The transport and reaction mechanisms inside the MEA are crucial for the performance of fuel cells. Hence, important for the efficient operation of fuel cells is the optimization of appropriate electrode structures [24]. There are various techniques available for fabricating MEAs, including the decal method, hot-pressing method, and spray-coating method (Fig. 5) [25].

2.3 *Role of Electrocatalysts in PEMWEs*

Catalysts play a crucial role in facilitating the electrochemical reactions at the cathode and anode of the PEMWE systems. It is reported that research is ongoing to develop cost-effective and efficient catalysts for PEMWE, as the use of precious metals like Pt can contribute significantly to the overall cost. Non-precious metal catalysts and other innovative materials are being explored to make PEMWE more economically viable for large-scale hydrogen production. The utilization of materials like graphite, Co, and Ni holds the potential to develop alternatives that could replace traditional electrocatalysts, offering lower costs and reduced environmental impact. Nonetheless, the performance advantages and limitations of these novel catalysts are pivotal considerations in determining the position of PEMWE technology in future energy conversion initiatives. In PEMWE systems, two primary reactions occur, namely HER and OER. HER reactions take place at the cathode, while OER reactions occur at the anode [26]. The rapid kinetics exhibited by the HER on the commonly employed Pt/C catalyst in the cathode side of PEMWE allow the loading to be reduced from 0.3 to 0.025 mg/cm^2 without compromising performance [27]. Over the past decade, several transition metal compounds, including borides [28], selenides [29, 30], sulfides [31], tellurides [32, 33], nitrides, and phosphides [34–36] have been compared to commercial Pt/C with their remarkable activity of HER in acidic electrolytes. While considerable research efforts have historically focused on tailoring composition, microstructure, catalyst/support interactions, and surface chemistry to enhance acidic HER performance, recent efforts have shifted towards minimizing the reliance on metallic constituents in catalysts. Moreover, there has been an emphasis on designing intricate hybrid catalysts encompassing multiple components, with the aim of synergistically enhancing overall catalytic performance. On the other hand, the OER process, which is characterized by its sluggish kinetics, significantly contributes to overall polarization and persists as a prominent research challenge [37]. In contrast to HER, the acidic OER process takes place under harsher and more stringent conditions. Therefore, the catalysts' electrochemical stability assumes paramount importance and poses a daunting challenge to be surmounted for the sustained and prolonged operation of PEMWE systems [38]. A correlation between the effectiveness and durability of oxides used as OER catalysts in acidic environments was found [39]. In particular, the most active oxides exhibit an explosion in dissolution rates by magnitude orders, making them the least stable [40]. As a result, IrO_2, which strikes a balance between stability and activity, emerges as the preferred catalyst in PEMWE systems, and IrO_2 accounts for approximately 8% of the total stack cost. This indicates that while IrO_2 is effective, it is also a relatively expensive component in the overall cost of the PEMWE stack. Balancing cost considerations with performance is crucial in the design and optimization of such systems. Therefore, researchers often work towards developing catalysts that maintain high activity and stability while reducing costs to make the technology more economically viable [41]. Ir makes up only a tiny fraction of the earth's atmosphere, making it one of the least abundant elements. Its production rate is contingent on the primary metal's extraction rate, as it is considered a secondary mining metal [42]. Therefore, it is imperative to reduce Ir loadings while enhancing OER performance. Furthermore, the slow kinetics of OER and the instability of materials used as OER electrocatalysts pose significant challenges to the advancement of PEMWE systems, rendering this issue a robust area of research [43]. According to recent projections based on Pourbaix diagrams, non-PGM elements such as Ti, Si, Co, Mn, Fe, Ag, Cu, Mo, Te, Sn, W, Ta, Pb, Bi, and Tl exhibit thermodynamically stabile oxides under acidic OER conditions. Consequently, these elements emerge as potential candidates for the development of novel PGM-free OER catalysts. It is noteworthy that metals or metal oxides exhibiting instability under acidic OER conditions are unlikely to find substantial utility in this regard [44].

3 Current State of Electrocatalysts in PEMWEs

3.1 Common Materials

In electrolyzers, electrocatalysts are selected according to various properties such as pore structure, particle size, high surface area, electrical conductivity, and corrosion stability [45]. Harsh electrochemical environments in PEMWEs, such as low pH, high anodic overpotential, strong oxidants present, and the ability to perform at higher temperatures, require the use of precious metal compounds in these systems [46]. In conventional PEMWEs, different catalysts are used for the HER and the OER. While Pt is generally used on the cathode side, Ir-based catalysts metal oxides are used at the anode [47]. As a support material, carbon doping is preferred to Pt in the cathode region. The details of currently used and commercially preferred HER and OER catalysts can be specified as follows:

3.1.1 Commercial HER Catalyst

The HER reaction, which is referred to as the Volmer step, starts with the adsorption of hydrogen on the electrocatalyst's surface. The Volmer step is crucial because it determines the rate of the overall HER. It involves the initial adsorption of hydrogen on the catalyst surface, and subsequent steps follow to complete the reaction. The efficiency and kinetics of the HER are essential considerations in designing electrocatalysts for applications like fuel cells and water electrolysis. The reaction can proceed as a Tafel reaction or Heyrovsky reaction to produce H_2. Since the adsorption of hydrogen is very important step, electrocatalysts are of great importance to facilitate this process. The H_2 adsorption free energy on Pt is almost thermoneutral ($\Delta GH^* \approx 0$). It is one of the most important reasons that Pt is an excellent choice for HER [48]. Although Pt catalysts are considered the best choice, they increase the cost of electrolyzers considerably. Various support additives are used to reduce these costs. Nowadays, carbon is preferred as an electrocatalytic support to decrease noble metal content and increase dispersibility [49]. However, carbon-supported Pt-based electrocatalysts are considered reference catalysts for HER [50]. The commercial HER catalyst was selected at 40 wt% Pt/C and Pt loading was reduced to 0.4–0.6 mg/cm^2. The use of Pt/C considerably reduced the Pt content and the overall costs [3]. Unfortunately, carbon used as a catalyst in OER mechanisms leads to catalyst shedding at the high anodic potential of OER. For this reason, it is only used as HER support in PEMWEs.

3.1.2 Commercial OER Catalyst

The OER mechanism in PEMWEs takes place in an acidic environment. The acidic environment is highly corrosive. Electrocatalysts must exhibit active and stable behavior in this corrosive environment. In addition, the content of more than one oxygen-containing intermediate (OH, O, OOH, etc.) depends on catalyst performance by optimizing the adsorption of active intermediates [3]. Many materials including noble metals and non-noble metals are investigated and Ir metal showed the best performance. Even though Ir exhibits stability and high activity, for being a very scarce material, reducing the amount of Ir in the catalyst layer is the main way to reduce the total cost. Commercially used IrO_2 catalyst loading in PEMWEs is currently in the range of 1–5 mg/cm^2 [6]. In addition to Ir metal, Ru metal is also preferred in the OER mechanism. When electrocatalysts are examined, Ru or RuO_2 is more active than Ir or IrO_2. However, the instability of Ru and its low resistance in corrosive environments limit its use in electrolyzers [51].

3.2 Efficiency and Resilience Concerns

PEMWEs are used as the preferred ideal power source for mobile and stationary systems. One of the most important factors determining the lifetime of PEMWEs in these systems is the catalyst. The choice of catalyst determines the degradation rate in an electrolytic cell. The lifetime of a PEMWE electrocatalyst is highly dependent on the catalysts. Catalysts used in both the anode and cathode sections of electrolyzer systems can undergo various deformations under long-term operating conditions, which leads to reduced cell efficiency. Under long-term operating conditions, difficulties such as corrosion of the catalyst, agglomeration of surfactants, dissolution of the catalyst surface, loss of activity of the active surface area, and exfoliation of the catalyst layer occur. All these challenges lead to a loss of structural integrity in the catalyst, which causes reduced efficiency and durability.

3.2.1 Corrosion and Degradation of Catalyst-Coated Membrane Assembly

The primary degradation mechanisms are the dissolution of the chemical decomposition of the membrane, anode catalyst, and formation of semiconducting oxides on metal components. The degradation speed is accelerated by dynamic operation, high current density, and shutdown modes. The main degradation problem is catalyst dissolution, leading to irreversible degradation of anode catalysts. The dissolution rate of IrO_x during the OER in the anode acidic region is related to the applied potential. As the applied potential increases, the amount of decomposition increases. Long working hours and acidic environment result in the loss of a large part of the catalyst. Ionomer loss and other structural changes can also be the cause of degradation [52]. All these reasons negatively affect the efficiency by reducing cell performance.

3.2.2 Agglomeration of Catalyst Surfactants and Catalyst Surface Dissolution

Agglomeration (growth of the crystal structure) and dissolution of the catalyst can lead to the formation of vacancies and ionomer dissociation. Catalyst agglomeration leads to a decrease in current density, but the high overpotential weakens the van der Waals bonds between the catalyst and the support structure, which leads to accelerated agglomeration. This ultimately leads to the collapse and dissolution of the electrocatalyst layer. These degradations are one of the most important parameters that reduce the efficiency and durability of the electrolyzer [53]. Agglomeration in catalysts reduces the activity of the surface area. Particle size, surface properties, crystal structure, and chemical structure can be among the parameters controlling the agglomeration behavior of the catalyst. Electrocatalyst layer morphology supports the effective use of the catalyst and determines the accessibility of the active region of the cell. For example, Pt nanotubes undergo dissolution at a lower rate compared to Pt nanoparticles. The electrode consisting of a nano-thin film has a lower agglomeration rate. The corrosion rate is also lower than in other materials due to the morphology of the catalyst [54]. The currently used Pt, Pt/C, Ir, and IrO_2 catalysts can exhibit high performance at low operating hours and low current densities; at high operating hours and high current densities, they can suffer from degradation and surface dissolution. This raises concerns about the efficiency and durability of PEMWEs. Solutions and research to improve the performance of these catalysts are ongoing.

3.3 Environmental and Economic Impacts

The economic and environmental impacts of PEMWE technology, especially concerning the utilization of rare metals like Ir and Pt, are considered. These metals constitute a significant portion of the system's cost. In particular, Ir metal is considered one of the most rare elements on the Earth's surface, accounting for only 1/10 of the Pt deposit [55]. However, due to the high cost of Pt, studies have shown that the cathode catalyst load can be significantly reduced. Today, the cathode loading is 0.2 to 0.5 $mg.cm^{-2}$. On the other hand, an Ir or its oxidized form catalyst is generally used on the anode side. However, the anode loading has not been significantly reduced due to overpotential since early development, and currently, the loading typically ranges from 2 to 3 mg.cm^{-2} [56]. If PEMWE systems are widely adopted on a global scale, this could lead to constraints in the supply of metals. In this context, various efforts are being made to enhance the sustainability of PEMWE technology and reduce costs. Efforts to reduce the reliance on precious metal catalyst materials or to replace them with more economical alternatives should be considered from the outset of the development process. Numerous studies have shown that transition metals (e.g., Ni, Mn, Fe, Cu, Mo, and Co) and carbon-based materials, which are plentiful in soils, demonstrate high performance for HER [57]. The utilization of these inexpensive electrocatalysts will significantly decrease the cost of PEMWE systems. It should be taken into account that the mentioned electroactive transition metals and carbon are abundant in common waste streams, like electronic waste, biowaste, and wastewater [58]. It is suggested that synthesizing electrocatalysts from waste materials could help achieve a 45% drop in carbon emissions by about 2030 and carbon neutrality by about 2050 [59]. The use of catalysts derived from waste significantly contributes to the sustainable development of green hydrogen energy, in line with the principles of the circular economy. This innovative approach aims to minimize environmental impacts by enhancing sustainability in waste management and energy production processes. The methods used to synthesize these catalysts include various techniques such as microwave, electrochemical synthesis, wet-chemical methods, electrochemical synthesis, and pyrolysis synthesis. Figure 6 shows a visual representation of the synthesis methods.

With the use of these techniques, various wastes, such as biological and electronic wastes, are converted into electrocatalysts. Recent literature studies will contribute to a sustainable energy future by aiming to minimize the environmental and economic impacts of PEMWE [60–62].

4 Challenges in Developing Sustainable Electrocatalysts

4.1 Eco-Designed Electrocatalysts

The processing, extraction, and purification of rare metals involve various challenging processes that can be environmentally harmful, such as hydrometallurgical methods [63]. The eco-designed electrocatalyst approach aims to replace environmentally harmful and negatively impacting methods by focusing on low carbon emissions and deriving from waste. Eco-design electrocatalysts can be categorized into biomass, e-waste, and natural minerals, as shown in Fig. 7 [64].

Biomass: Biomass derived from sources such as food waste, animals, plants, and microorganisms can be utilized to construct carbon-based electrocatalysts. These renewable materials offer an eco-friendly alternative and contribute to reducing carbon emissions.

Electronic Waste (E-waste): E-waste, comprising discarded electronic devices, can be a valuable source of metals for electrocatalyst production. By recycling

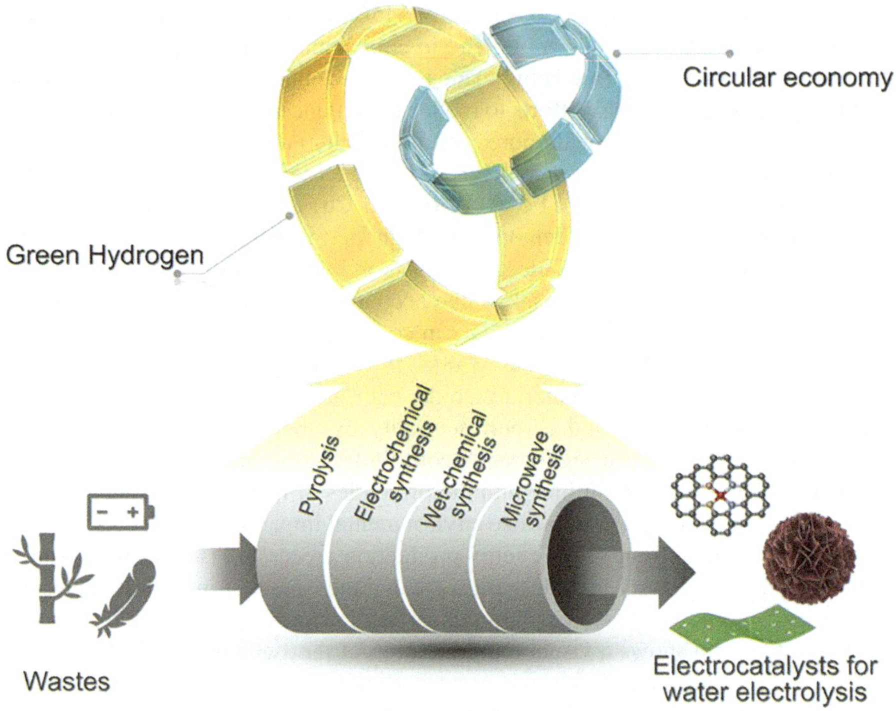

Fig. 6 Schematic of green hydrogen energy based on circular economy supported by waste-derived electrocatalysts for water electrolysis. (Reprinted with permission from ref. 58. Copyright 2023, Springer [58])

Fig. 7 Eco-design electrocatalyst

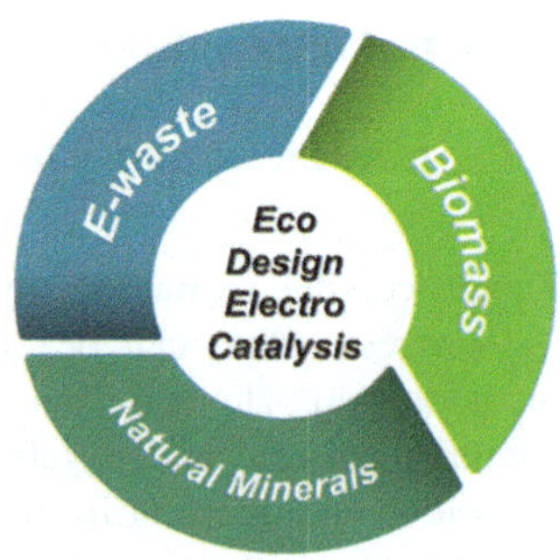

e-waste, precious metals can be recovered and repurposed, reducing the need for primary metal extraction and minimizing environmental harm.

Natural Minerals: Certain natural minerals, such as Fe, Ni, Co, Mn, Cu, and Mo-based oxides and sulfides, can be used in electrocatalyst design. These minerals provide abundant and sustainable options for catalytic materials while minimizing the negative environmental impact associated with traditional methods.

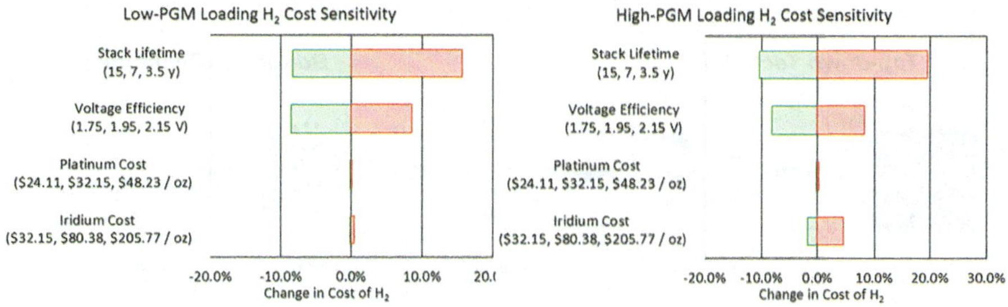

Fig. 8 H_2 cost (OpEx and CapEx included) with variations in voltage efficiency, lifetime, Ir cost, and Pt cost. Left: Low PGM loading of 0.05 mg Pt/cm^2 and 0.25 mg Ir/cm^2. Right: High PGM loading 1.0 mg Pt/cm^2 and 3.0 mg Ir/cm^2. (Reprinted with permission from ref. 70. Copyright 2022, American Chemical Society [70])

4.2 *Catalyst Degradation and Lifetime*

Several factors affect catalyst degradation and lifespan, such as temperature, pH, overpotential, and other effects. These influences can lead to a decrease in catalyst degradation and reaction kinetics [65]. Dissolution, leaching, Ostwald ripening, coalescence, phase transition mechanisms, and segregation, are factors that can affect catalyst degradation and lifespan. These mechanisms can lead to changes in the catalyst structure, composition, and morphology, resulting in a decrease in its performance and reaction kinetics over time [66]. The Accelerated Stress Test (AST) is a test protocol used to understand the degradation and long-life effects of PEMWE. In this process, the electrolyzer is subjected to testing for more than 1000 h under high current density (>1 A cm^2) conditions. This test can result in catalyst dissolution formation and losses in cell performance due to degradation effects. Moreover, this test provides information about the long-term durability of the electrolyzer cell [67]. For example, Milosevic et al. [68] showed that degradation in lifetime prediction studies for PEMWE, especially the dissolution and transport of anode catalyst Ir into the MEA layer, significantly contributes to catalyst degradation and negatively affects the lifetime of the PEMWE system. Burch et al. [69] investigated the effects of Ir-coated polymer membrane thickness in PEMWE during tests exceeding 5000 h. They examined the exponential change in Ir dissolution with operating time and demonstrated a decrease in cell performance in the tests. Efforts to improve the lifetime of PEMWEs involve developing more stable catalyst materials, optimizing operating conditions, and designing systems with enhanced durability. Regular maintenance and careful management of operating conditions can help mitigate catalyst degradation and extend the overall lifetime of PEMWE. Researchers are exploring alternative catalyst materials to decrease costs and increase the resilience of the electrolyzer components such as non-precious metals. Moreover, advancements in materials science are continually contributing to the development of more robust and durable electrolyzer technologies.

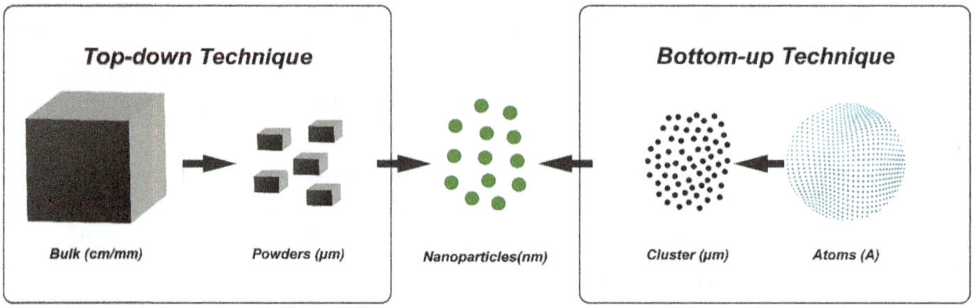

Fig. 9 Top-down and bottom-up techniques. (Reproduced with permission from ref. 71. Copyright 2023, MDPI [71])

4.3 Scale-Up and Manufacturing Challenge

One of the factors that affects hydrogen generation costs is the use of PGM like Pt and Pd as catalysts. Although these metals have high costs, they have an important position in energy conversion applications due to their high activities. In Fig. 8, two different catalyst loads were examined for voltage efficiency, stack life, Pt cost, and Ir cost. Catalyst loadings prepared to be used in the anode and cathode have been prepared in two different types: low (0.05 mg Pt/cm^2 and 0.25 mg Ir/cm^2) and high loading (1.0 mg Pt/cm^2 and 3.0 mg Ir/cm^2). Analysis showed that Ir costs decreased significantly at low loads [70].

5 Innovative Approaches to Sustainable Electrocatalysts

Nanotechnology in Electrocatalyst Development

Nanoparticle synthesis is divided into two primary methods: top-down and bottom-up, as shown in Fig. 9.

The top-down method involves reducing the size of a solid material or bulk structure through a series of physical processes. Tools such as grinders, ball mills, and laser lithography are used in the size reduction stage. Moreover, the bottom-up approach involves the chemical synthesis of the material. It includes factors such as pH, bonding, temperature, and concentration [72].

5.1 Top-Down

5.1.1 Nanomaterial Synthesis via Mechanical Milling

The ball milling process is a mechanical grinding process that reduces the grain size and alters the surface properties of materials. Particles undergo compression, leading to structural changes. This process is utilized for obtaining materials in the nanoscale range, but it can also result in particle agglomeration. One of the advantages of ball milling is its ability to accommodate large-capacity processes and enable high-volume production. It is a safe, simple, and long-term applicable technique [73, 74].

5.1.2 Laser Ablation

Laser ablation, which can be a non-thermal or thermal process, is the removal of atoms from a solid material using a focused pulsed continuous wave or laser beam. This technique is used for various applications, including the production of thin films and nanoparticles, as well as drilling small holes in hard materials. Laser ablation can be used to remove metal parts that do not volatilize upon heating, or metal laminates with varying gas-phase properties. Carbon nanotubes, fullerenes, and metals are examples of products that can be obtained through this process [75]. Laser ablation involves directing a laser beam at a solid target material. The beam releases atoms, molecules, and ions to form a plasma plume. The plasma cloud grows to supersaturation, and then fast cooling leads to condensation nucleation. 1D structures like nanowires and nanotubes can also form during the deposition process. It is important to note that laser ablation can be performed in a liquid phase, resulting in the production of even smaller particles. This liquid-phase ablation is particularly suitable for generating nanoparticles with complex compositions, as the stoichiometry of elements is not affected by the process [76].

5.1.3 Nanosphere Lithography (NSL)

NSL is a cost-effective and high-throughput method that is widely used in laboratories worldwide. One of the primary benefits of NSL is its ability to create homogeneous arrays of nanoparticles, which can be utilized in various studies. These nanoparticles can be used to investigate the size-dependent magnetic, optical, catalytic, thermodynamic, electrical, and electrochemical transport properties of materials. However, NSL has a few perceived limitations. First, it is only able to produce a limited range of particle distances and nanoparticle shapes. Second, there is a limitation in size due to increasing polydispersities in the nanosphere synthesis. In order to grow a 2D colloidal crystal in the 10–100 nm range, the nanospheres have to be of smaller diameter [77]. The NSL process conventionally requires a nanosphere mask to be self-assembled on a substrate, and then materials are deposited in line of sight across the mask. The substrate is usually placed perpendicular to the deposition direction. The resulting nanoparticles are in-plane in shape and spacing, as defined by the projection of the nanosphere mask onto the substrate. NSL is an attractive nanofabrication technique due to its low cost, high throughput, and broad applicability. While it has limitations in terms of nanoparticle shapes and size ranges, it is still widely used in research laboratories for various material studies [78].

5.2 Bottom-Up

In contrast to top-down methods, bottom-up approaches offer the advantage of creating nanostructures with fewer defects. These methods utilize nanoparticles, molecules, and atoms as the starting materials to construct complex structures. Ionic and molecular self-assembly is an approach that relies on physical or chemical forces to assemble individual building blocks or molecules into larger structures through noncovalent bonds, like ionic and hydrogen bonds, and van der Waals forces. Bottom-up methods provide the ability to precisely control the spatial properties and composition of the individual building blocks, allowing for tailored nanostructures [79]. The size of the building blocks can be adjusted according to the required features. Overall, bottom-up approaches offer advantages in terms of creating nanostructures with fewer faults compared to top-down methods. They leverage the self-assembly of nanoparticles, molecules, and atoms to construct complex and precisely controlled nanostructures, enabling advancements in various fields of nanotechnology [80, 81].

5.2.1 Sol-Gel Method

Sol-Gel is a popular chemical approach to synthesizing Metal Oxide Nanoparticles (MONPs). It involves the conversion of metal alkoxides or metal salts through hydrolysis and subsequent condensation to form MONPs. The sol-gel process typically consists of three stages: hydrolysis, condensation, and aging. In the hydrolysis stage, a solution of metal alkoxides or metal salts is subjected to hydrolysis in a solvent medium. The hydrolysis reaction converts the alkoxides or salts into metal hydroxides, creating a sol-like solution. During the condensation stage, the metal hydroxides undergo further reactions, such as polycondensation, leading to the formation of three-dimensional networks. These networks grow and develop into a gel, which is a three-dimensional solid-like structure containing a solvent [82]. The aging stage involves allowing the gel to mature or age under controlled conditions, such as specific temperature and humidity. This aging process facilitates the consolidation and structural rearrangement of the gel network, leading to the formation of more stable and well-defined MONPs. The sol-gel method offers several advantages for MONP synthesis. It allows accurate control of the composition, shape, size, and structure of the NPs. The sol-gel method is a chemical approach that involves the hydrolysis and condensation of metal alkoxides or metal salts to produce metal oxide nanoparticles [83].

5.2.2 Co-precipitation

Co-precipitation is a common method for the synthesis of nanoparticles such as magnetite (Fe_3O_4). Simultaneous precipitation of ferrous and ferric ions occurs in a basic solution. However, it can be challenging to control the particle size and size distribution using this method. Additional treatments or size selection techniques may be necessary to achieve the desired size range. Surfactants can be added to the reaction media or used for coating to produce well-dispersed magnetic nanoparticles [84]. They help regulate particle size and stabilize colloidal dispersions. Despite improvements in aqueous synthesis approaches for magnetite nanoparticles, controlling particle size remains a significant challenge [85].

5.2.3 Hydrothermal Synthesis

The hydrothermal method is the widely used method for the preparation of nanomaterials through solution-based reactions. It involves the formation of nanomaterials over a wide temperature range. By controlling the pressure conditions, either high or low, the morphology of the synthesized materials can be controlled based on the steam pressure of the major reaction composition. Hydrothermal synthesis offers several advantages over other methods. It allows for the synthesis of nanomaterials that are not resistant to higher temperatures and minimal loss of materials. Solvothermal and hydrothermal synthesis methods are promising approaches for preparing nanomaterials because of their advantages such as cost effectiveness, high crystallinity of nanocrystals, and control over dimensions. These methods can be integrated with microwave and magnetic field techniques to achieve semi-continuous material synthesis with enhanced high quality and reproducibility [86, 87].

Advances in Computational Modeling for Catalyst Design

Density Functional Theory (DFT) is a computational method used in electrocatalysis applications. It is based on the Schrödinger Equation of quantum mechanics, which describes the behavior of a particle. DFT incorporates the equations of quantum mechanics and provides predictions about the relationship between the chemical and physical properties of an electrocatalyst and its electrochemical performance [88]. In today's catalyst design applications, trial-and-error methods are used to develop catalysts that exhibit high activity and stability. DFT calculations enable us to predict complex systems, allowing us to gain insights into the activity and other catalytic effects of catalysts used in electrocatalysis applications such as the HER and OER mechanisms of PEMWE, without the need for synthesizing different recombination variations of these catalysts [89]. DFT calculations include a gas phase correction for OER calculations, and gas phase corrections account for the interactions between reactants and products, thereby providing the necessary adjustments. Gas phase corrections are one of the areas that need improvement in DFT applications, and accurate modeling is essential for correct calculations. If the corrections are not made appropriately, critical parameters such as activity and reaction energies are calculated incorrectly, leading to inaccurate predictions about the catalyst's actual performance [90].

6 Emerging Materials for Sustainable Electrocatalysts

The development and applications of inexpensive, readily available, and robust non-precious metal catalysts as alternatives to Pt-based catalysts are extremely important for sustainable PEMWE [91]. In recent years, there have been significant attempts to achieve stabilization of non-precious metal catalysts in acidic environments [92–95]. Some metals with strong hydrogen-binding capabilities, such as Fe, Ni, and Cu, are electrochemically employed for coating electrodes, including binary and ternary alloys. These alloys serve as highly effective electrocatalysts for HER, and the combination of two or more metals can improve HER performance [96]. Various metals have been theoretically and experimentally investigated for their OER and HER performances [1, 97, 98]. For example, the active contributions of non-precious metal oxides like TiO_2, Ta_2O_5, SnO_2, Sb_2O_5, and Nb_2O_5 to the OER have been the subject of significant research for sustainable PEMWE [99, 100]. The need to reduce or replace the use of Ir with other non-precious metal oxides has been emphasized. Therefore, substantial research is highly advantageous for developing low-cost electrocatalysts for OER, characterized by durability and large surface areas [101]. Carbon materials like carbon nanotubes, carbon nanosheets, and graphene are widely utilized in the catalytic electrodes of PEMWE [102]. The high surface area of carbon-based materials provides more catalytically active sites, allowing for more efficient electrochemical reactions [103]. However, the role of carbon-based materials in PEMWE is not limited to catalysis alone. Moreover, their durability and conductivity can enhance the total performance and lifespan of the electrolyzer. Specifically, the use of carbon nanotubes in membrane components can increase proton conductivity, thereby improving electrolysis efficiency [104, 105]. Xe et al. [106] hybridized carbon nanotubes with tungstophosphoric acid (PWA) heteropolyanions and tested them for the HER. The new catalyst alternative was found to offer activity up to 20% of Pt/carbon nanotube activity. WO_3 nanorods for HER were also demonstrated, and they indicated the potential for high-efficiency production using a single-step method. To replace precious metal catalysts in commercial PEMWE, non-precious metals need to achieve high current densities under acidic conditions [107]. Non-precious metals are commonly preferred materials for PEMWE applications, and composite materials are obtained by combining them with materials possessing excellent properties, such as carbon nanotubes [1]. Composite materials are formed by combining materials with different properties. For instance, polymer matrix composites reinforced with carbon nanotubes can increase the electrocatalytic active surface area and extend the electrode's lifespan by enhancing mechanical durability [108]. Also, combinations of materials such as graphene and titanium dioxide can be optimized for their performance in electrocatalysis [109–111]. These advancements contribute to making PEMWE technology more appealing from a sustainability perspective. The use of non-precious metals, carbon-based materials, as well as composite and hybrid materials, has the potential to enhance energy conversion efficiency while simultaneously reducing production costs. However, further research and development in this field, particularly in material synthesis, characterization, and scaling, can accelerate the widespread adoption of these materials in PEMWE technology. The discovery and assessment of catalysts for PEMWE that do not contain precious metals are crucial for realizing a hydrogen-based energy economy. In a study by King et al. [112] the sustainability of terawatt-scale PEMWE was explored using cobalt phosphide (CoP). The study found that CoP, when used in the HER, provided significant material cost savings compared to Pt, operating effectively and stably for over 1700 hours at 1.86 $A.cm^{-2}$ for continuous hydrogen production. In another study aimed at overcoming the unsustainable nature of Pt-based electrocatalysts, Giovanni et al. [113] evaluated Fe_xS_y catalysts in PEMWE single-cell tests. They discovered that pyrite (FeS_2) had the capacity to achieve 2 $A.cm^{-2}$ current density @ 2.3 V, providing an inexpensive and durable transition metal catalyst for the hydrogen formation reaction. Kelly et al. [114] assessed Mo_2C and Pd supported on WC for HER activity, demonstrating their stability and cost-effectiveness as HER catalysts. The use of non-precious metal catalysts also has a significant impact on sustainability in the context of OER. Bunea et al. [115] reported that antimony-doped tin oxide (SnO_2) served as a stable OER catalyst support, minimizing ohmic losses while providing sufficient electronic conductivity. In their single-cell tests, they showed that catalysts with up to 20% tin content exhibited improved cell performance compared to MEAs with a similar Ir black loading, highlighting the enhanced cell performance of catalysts with higher tin content. Developing highly and modern efficient catalysts for the HER and OER processes presents researchers and scientists with numerous innovative research opportunities. There is still a need for catalysts with promising current density by the industry. Further research is required for non-precious catalysts in the PEMWE process [50, 116].

7 Environmental and Economic Implications

The development of catalysts can provide a significant decrease in energy consumption, environmental pollution, and costs for electrochemical processes. For instance, the automotive industry's introduction of the three-way catalyst (Pt, Rh, and Pd) has prevented over 1 billion tons of emissions such as NO_x and CO from reaching the atmosphere. On the other hand, in addition to the production of materials for hydrogen storage and photovoltaic cells (PVCs), catalysts play a pivotal role in other sustainable processes like CO_2 capture and conversion, as well as the processes of water electrolysis and fuel cells [117, 118]. The development of alternative materials contributes to the sustainability of the hydrogen production process to ensure less dependence on scarce resources. For hydrogen to become a widespread clean energy source, scalable and cost-effective production methods are needed. Minimizing the use of expensive materials enhances the scalability of PEMWE technology. Therefore, researchers are actively working on developing alternative catalysts and materials that can replace or reduce the dependence on these precious metals. It also includes exploring non-noble metal catalysts and optimizing the composition of electrodes to maintain active performance while reducing costs. Advances in materials science, nanotechnology, and catalyst design are helping to form more efficient and durable PEMWEs with lower noble metal content. This is important to make green hydrogen production economically viable and competitive with other energy sources [119]. The sustainability of energy conversion devices must be assessed using a comprehensive approach that considers environmental impacts and cumulated energy factors, along with eco-design strategies. FCs are accepted as clean technologies because of lower emissions during the operation stage compared to conventional energy conversion systems, but components of FCs must be analyzed from manufacturing to the disposal phase to reveal the potential environmental impacts. Electrocatalysts in a typical electrolyzer system are determined as the main environmental contributors, and analysis of potential life-cycle impacts is crucial for understanding the environmental implications of sustainable electrocatalyst materials. Electrocatalysts in a typical electrolyzer system are determined as the main environmental contributors up to 78% global warming potential (GWP) (kg CO_{2eq}) compared to other components [120]. Therefore, analysis of potential life-cycle impacts is crucial for understanding the environmental implications of sustainable electrocatalyst materials. Extraction, development, and disposal of electrocatalysts can lead to significant impacts, such as cumulative energy demand (CED), carbon footprint, and waste production throughout their lifetime [121]. Life cycle assessment (LCA), including extraction and processing of raw materials, use phase, end-of-life disposal, and manufacturing, is a holistic tool to evaluate the overall sustainability and identify areas for improvement points for electrocatalyst [122]. LCA methodology of electrocatalysts depends on the goal and scope definition, boundary, and lifetime (operating hour) of the considered system, functional unit, life cycle inventory (LCI) data, selected impact factors, and target audience as well as four standard steps of a typical LCA study. The quality of LCI data, including input material and energy flows, is the primary factor affecting the results of the analysis and is based on completeness, temporal and geographic coverage, and consistency [123]. Developments in extraction, mining, and metallurgical processes of electrocatalysts, such as Pt, Pd, and Rh, mitigate the environmental impacts substantially, such as 95% recycled Pt mass input yields 29% decrease in GWP for manufacturing a $48kW_{el}$ PEMFC stack [121]. Reducing the Pt content in the manufacturing process of FC stack can lead to a substantial decrease in environmental impacts, in particular GWP arising from degraded stack replacement during the operation stage [124], and recovering the electrocatalysts at the end-of-life disposal stage prevents up to half of the total environmental impacts [125]. Moreover, analyzing the cumulative energy demand values of electrocatalysts reveals the primary energy consumption rates and advances in calculating the energy payback times of fuel cells. Environmental and energy impact analysis of electrocatalysts provides comprehensive guidance for developing sustainable and environmentally friendly materials to ensure energy conversion devices are entirely clean power generation technologies [126, 127]. By improving the selectivity and efficiency of catalysts, it becomes possible to minimize waste and energy consumption in chemical manufacturing processes. These contribute to reducing the carbon footprint associated with the production of these chemicals, aligning such improvements with broader efforts to transition to more sustainable and environmentally friendly industrial practices. To advance decarbonization and pave the way for global economic opportunities across various industrial sectors, low-cost green hydrogen production, storage, utilization, and transportation need to be developed. Developing a robust infrastructure for transporting and distributing hydrogen, and therefore pipelines, shipping, and trucking that can ensure the efficient and safe delivery of hydrogen to end users, is important [128]. It is critical to identify and encourage sectors where green hydrogen obtained by using PEMWEs can be effectively utilized. Industries like manufacturing, transportation, and energy production can benefit from the integration of green hydrogen, helping them reduce their carbon footprint for zero emissions [129]. The forthcoming challenges that will arise in the coming period regarding the investment costs of electrolysis of water may significantly affect the hydrogen-based energy system. The prospect of achieving low investment costs can provide significant benefits to both consumers and plant operators, exerting a positive impact on overall hydrogen pricing dynamics [130]. Governments and international organizations play an important role in providing policy frameworks, incentives, and regulations that encourage the growth of green hydrogen. A sustainable energy industry relying on hydrogen is currently being gradually implemented by society. Because supportive policies can accelerate investment, research, and development, especially in the energy sectors. To accelerate this transition, both national and international efforts are underway, aiming to lay the foundation for the emergence of hydrogen as the primary energy carrier in the future [5].

8 Future Directions and Research Needs

8.1 Collaborations

In May 2022, the European Union (EU) Commission identified clean energy production, energy conservation, and diversification of energy supply as priority areas. Emphasizing that hydrogen is a key element for these strategic areas, the Commission aims to increase the capacity of water electrolyzers to 17.5 GW by 2025, producing ten million tons of hydrogen per year [131]. Similarly, the UK has increased its 2030 targets from 5 to 10 GW, in line with its hydrogen priority areas. Australia, which has a high renewable energy potential, announced the Australian National Hydrogen Strategy in 2019 and has identified being one of at least three hydrogen suppliers in the Asian market as a success criterion [132]. In this context, the global promotion of hydrogen brings with it a variety of policy measures, collaborations, and financing opportunities. The opportunities for cooperation between countries, industrial producers, universities, and research institutions can be summarized as follows. Industrial manufacturers are entering into various cooperation agreements to bring electrolyzer systems into widespread use. MoUs signed between leading companies in the sector can enable them to lead the green hydrogen era. These collaborations can enable the expansion of not only local but also international green hydrogen initiatives and explore new business opportunities to increase the competitiveness of the green hydrogen industry [133]. In addition to industrial collaborations, another collaboration opportunity is international collaborations. Countries aim to create a market for hydrogen energy and electrolyzer systems through investment and R&D programs. For example, in November 2022, the Organization for Economic Cooperation and Development (OECD) launched the "Green hydrogen opportunities for emerging and developing economies: Identifying success factors and creating enabling conditions for market development." They then organized seminars with countries such as Colombia, Egypt, India, Indonesia, the Philippines, and Thailand. In these seminars, the needs of the countries were discussed. These inter-country studies provide the role that the private sector and international organizations can play and facilitate international hydrogen trade [134]. With the increase in such studies and collaborations in the future, the demand for green hydrogen systems will increase at the same rate. International projects that have taken place and will take place will be carried out by universities and research institutions, making the use of hydrogen easier and overcoming the difficulties of use.

8.2 Financing Opportunities

The increase in green hydrogen projects will be made possible by the availability of collaborating companies and appropriate capital. International organizations like IRENA and the GCF can help accelerate the transfer of financial resources and technological know-how between countries, resulting in the establishment of a technology infrastructure for clean hydrogen [135, 136]. According to a Bloomberg report, as part of the International Energy Agency's (IEA) efforts to achieve its "net zero emissions by 2050" target, approximately USD 13 million of private sector investment is needed in the hydrogen and carbon capture, utilization, and storage (CCUS) sectors [137]. This investment represents a comprehensive financial commitment to sustainable energy projects and is critical for the development of clean energy technologies. Investments in green hydrogen at the global level are growing rapidly as countries and industrial firms begin to realize and mobilize more of its potential in the low-carbon transition. This is accelerating the green energy transition, providing not only economic benefits but also supporting environmental sustainability. This significant increase in financing and commitments shows that the energy sector is undergoing a strategic transformation and that the private sector has increased its confidence in clean energy projects. In this context, this positive trend among the international business community and financial institutions creates a strong base for the acceleration of the clean energy transition and the global energy transformation. Many countries have already announced significant investment plans for the hydrogen sector in the coming years. For example, the Government of Japan has set a target of six times the annual hydrogen supply by 2040, reaching

12 million tons. To support this ambitious goal, Japan has allocated a total of 15 trillion yen ($107 billion) in private and public resources for research and development and production [138]. The United States government has made $9.5 billion available for hydrogen energy funding through legislation approved in 2021. Likewise, the government of Germany has allocated $9.5 billion as part of its national hydrogen strategy. This budget specifically includes research and development and prototype development projects. The European Union is making significant investments with a focus on the development of sustainable hydrogen. The Horizon Europe program provides $1.09 billion for research and development and market development projects. Another unit of the European Union has also allocated a significant portion of its $891 billion budget for the advancement of hydrogen energy [139]. These global examples represent important steps in the growth of the hydrogen sector. They will foster greater collaboration between government and the private sector, strengthen financing partnerships for hydrogen projects, and open up new opportunities for investors looking to invest in the hydrogen economy. Investments realized in this way will boost confidence in the hydrogen sector, leading to its acceleration.

8.3 Research Gaps and Potential Paradigm Shifts on the Horizon

The high costs and low abundance of PGM based electrocatalysts have limited the widespread adoption of advanced electrochemical technologies. Research and development efforts in the field of PEM fuel cells have been focused on reducing the amount of Pt catalyst or using non-PGM-based catalysts. PEMWEs face significant challenges, particularly due to their dependence on scarce Ir elements. However, the discovery of Ir-free anode catalysts could support the broader-scale utilization of PEMWEs and eliminate the dependence on Ir [70]. Consequently, a major concern for large-scale PEMWE applications is the supply. A subject requiring further analysis and research is effectively addressing the durability issues encountered with low catalyst loadings. In this context, it is important for future studies to focus on developing strategies to enhance the long-term performance of low-loading electrocatalysts [140]. The anticipated deployment of PEMWEs from gigawatt to terawatt scale raises questions regarding the availability of PGM-based catalysts, particularly Pt and Ir. While there is an understanding of having sufficient PGM for electrolysis at the gigawatt scale, ensuring an adequate supply of PGM for the deployment of TW scale electrolyzers worldwide is a crucial question. However, Pt and Ir used in electrode production for PEMWEs are not the sole components responsible for high capital costs. Other equipment like BPs, GDLs, membranes, and stack BOP still contribute significantly to the expensive cost [70]. Therefore, it is imperative to concentrate on developing new techniques for on-site assessment of catalyst performance and durability. This will enable the creation of new generations of electrochemical catalysts that do not rely on scarce PGM-based elements. However, there are ongoing challenges in hydrogen production with affordable non-precious catalysts. These challenges include the inefficient properties of transition metal oxides in PEMWE, potential decline in performance due to overloading of non-precious metals on the anode, complex synthesis procedures for bifunctional catalysis, the synthesis process influencing the structure and performance of the catalyst, low electrical conductivity of transition metals, and the difficulty in preparing non-precious catalysts that can surpass the properties of precious metals [50]. To address the research gaps mentioned above, recent studies in the literature have been reviewed. In a study by Islam et al. [141], the role of alternative catalysts excluding PGMs in hydrogen production is explored. Among the supports investigated in this article, it was found that titanium dioxide support with macropores and low surface area (TiO_2-MC) is effective support for the Ir catalyst (Ir/TiO_2-MC). Ir/TiO_2-MC demonstrated remarkable performance in PEMWE with a low Ir loading of 0.27 mg $Ir.cm^{-2}$, outperforming even two types of commercial catalysts, achieving a notable performance of 1.75 $A.cm^{-2}$ @ 1.8 V. Another recent study by Wang et al. [142] presents significant findings aimed at developing an effective anode catalyst for PEMWE. These catalysts have managed to surpass commercial Ir catalysts significantly by utilizing metal oxides such as antimony-doped tin oxide (ATO) and Nb_2O_{5-x}. The achieved high performance and stability are attributed to the unique chemical interaction between Ir and Nb_2O_{5-x}. These results offer a promising strategy for the development of low Ir-loaded PEMWEs. While the need for non-precious metals may not be as critical in anodes as in cathodes, it remains an important factor. The absence of PGMs, particularly in the electronic and energy sectors, can impact production processes and reduce efficiency. Therefore, emphasizing the role of PGMs in cathodes underscores their critical importance in industrial production. These findings highlight the potential of non-PGM catalysts in PEMWEs. The high performance and stability of alternative catalysts represent a promising step towards reducing PGM usage in industry and adopting a more sustainable approach. These studies could play an important foundation for designing more effective and economically advantageous PEMWEs in the future. In this context, researching and developing alternative catalysts to replace PGMs in industrial applications represent significant strides towards a sustainable energy future.

9 Conclusion

The improvement of electrocatalysts or catalysts for producing hydrogen has become a focus of research because of the potential of hydrogen as a versatile and clean energy source for various applications, including transportation, industry, and power generation. By using abundant and eco-friendly materials, such as certain metals or metal oxides, researchers can design catalysts that are both effective and environmentally friendly. Electrocatalysts or catalysts are essential to facilitate electrochemical reactions, especially in processes such as electrolysis and fuel cell applications. These technologies use precious metals like Pt and Pd as catalysts, which can be expensive and environmentally effective. The fact that these metals are rare in nature and costly provides a significant disadvantage. Several materials have been investigated as potential electrocatalysts for PEMWEs, including PGMs, transition metal oxides, and carbon-based materials. On the other hand, PGMs, such as Pt and Ir, are expensive and rare, limiting their scalability and sustainability. The fact that these metals are rare in nature and costly provides a significant disadvantage. Carbon-based materials can be functionalized or doped with heteroatoms to improve their catalytic activity for water electrolysis. These materials can provide good catalytic activity for both HER and OER and can be more cost-effective alternatives to PGMs. However, sustainable catalysts are essential in the development of clean energy technologies, improving overall performance and reducing environmental impacts associated with traditional energy production methods. To evaluate the environmental impact of catalysts, this approach helps identify opportunities for improvement and ensures the overall sustainability of catalytic processes. The materials used in the catalyst should be environmentally friendly and non-toxic, both during production and operation. This is important for ensuring the overall sustainability of the technology. Moreover, the synthesis step or methods of electrocatalysts must be scalable to industrial production levels, especially to meet the widespread demands of PEMWE technology. Catalysts are often based on abundant and eco-friendly materials, contributing to efficient and cost-effective hydrogen generation. These catalysts support the integration of renewable or alternative energy technologies into current energy systems, and they enable the production of clean energy carriers such as hydrogen without the release of harmful emissions. They also promote the widespread adoption of clean energy solutions by enabling the storage and conversion of renewable resources into versatile forms of energy. Moreover, advancements in nanotechnology and material science have enabled the superior catalyst design with enhanced surface areas and catalytic activities. Catalysts play a role in converting alternative energy sources like solar energy or wind energy into storable and transportable forms like hydrogen. When powered by renewable energy sources, this process can contribute to achieving carbon neutrality by avoiding the use of fossil fuels for energy production. This is important for creating a sustainable and scalable energy infrastructure that minimizes the environmental footprint associated with clean energy production. Therefore, researchers must give priority to the design and development of catalysts based on green energy principles. This ensures minimizing the use of hazardous substances, reducing waste, and promoting energy efficiency in catalytic processes. Moreover, the adoption of sustainable catalysts can lead to the creation of new industries and business opportunities. With the growing need for clean energy technologies, the development of efficient catalysts is critical to economic growth and competitiveness. Sustainable catalysts for PEMWEs play a crucial role in advancing the technology toward commercialization by improving efficiency, reducing costs, and ensuring environmental sustainability. Ongoing research and development in this area are essential for realizing the full potential of PEMWEs for hydrogen production.

Acknowledgments This research is supported by the Scientific and Technological Research Council of Turkey (TUBITAK) 1001 Research Projects Funding Program under the project number:120M234. The authors are very grateful to the Scientific Research Projects Unit of Erciyes University for financial support (Project No: FYL-2023-12877). In addition, the second author MK thanks the Scientific and Technological Research Council of Turkey (TUBITAK) for their scholarships under the "2210-C Priority Areas Ph.D. Scholarship Program (grant number 1649B032309360)." M.K. also thanks the Turkish Higher Education Institution YÖK for their 100/2000 Ph.D. Scholarship Program. Authors E.Ö. (grant number 1649B022315393), M.C. (grant number 1649B022307820), and M.A.D. (grant number 1649B022305241) thank the Scientific and Technological Research Council of Turkey (TUBITAK) for their scholarship under the "2210-C National Priority Areas M.Sc. Scholarships Program."

The authors dedicated this publication to 100th anniversary of the Republic of Türkiye.

References

1. Kumar SS, Himabindu V (2019) Hydrogen production by PEM water electrolysis–a review. Mater Sci Energy Technol 2(3):442–454
2. Ajanovic A, Sayer M, Haas R (2022) The economics and the environmental benignity of different colors of hydrogen. Int J Hydrog Energy 47(57):24136–24154
3. Zhang K, Liang X, Wang L, Sun K, Wang Y, Xie Z et al (2022) Status and perspectives of key materials for PEM electrolyzer. Nano Res Energy 1(3):e9120032
4. Wang T, Cao X, Jiao L (2022) PEM water electrolysis for hydrogen production: fundamentals, advances, and prospects. Carbon Neutrality 1(1):21
5. Chatenet M, Pollet BG, Dekel DR, Dionigi F, Deseure J, Millet P et al (2022) Water electrolysis: from textbook knowledge to the latest scientific strategies and industrial developments. Chem Soc Rev 51(11):4583–4762
6. Ahmed KW, Jang MJ, Park MG, Chen Z, Fowler M (2022) Effect of components and operating conditions on the performance of PEM electrolyzers: a review. Electrochem 3(4): 581–612
7. Chen Y, Liu C, Xu J, Xia C, Wang P, Xia BY et al (2023) Key components and design strategy for a proton exchange membrane water electrolyzer. Small Struct 4(6):2200130
8. Wang Q, Li K, Wang Y, Chen H, Duan Q, Sun J (2018) The efficiency of dodecafluoro-2-methylpentan-3-one on suppressing the lithium ion battery fire. J Electrochem Energy Convers Storage 15(4):041001
9. Perović K, Morović S, Jukić A, Košutić K (2023) Alternative to conventional solutions in the development of membranes and hydrogen evolution electrocatalysts for application in proton exchange membrane water electrolysis: a review. Materials 16(18):6319
10. Zhang S, Wang Z, Zhang R, He Y, Cen K (2023) Comprehensive study and optimization of membrane electrode assembly structural composition in proton exchange membrane water electrolyzer. Int J Hydrog Energy. https://doi.org/10.1016/j.ijhydene.2023.05.280
11. Kibsgaard J, Chorkendorff I (2019) Considerations for the scaling-up of water splitting catalysts. Nat Energy 4(6):430–433
12. Ursache A (2023) Critical materials raw as leverage in global geopolitics. SCIENTIA MORALITAS-Int J Multidiscip Res 8(2):239–250
13. David M, Lyth SM, Lindner R, Harrington GF (2021) Future-proofing fuel cells: critical raw material governance in sustainable energy. Springer Nature, Cham
14. Guo Y, Li G, Zhou J, Liu Y (2019) Comparison between hydrogen production by alkaline water electrolysis and hydrogen production by PEM electrolysis. IOP Conf Ser Earth Environ Sci 371:042022. IOP Publishing

15. Falcão D, Pinto A (2020) A review on PEM electrolyzer modelling: guidelines for beginners. J Clean Prod 261:121184
16. Wilberforce T, El Hassan Z, Ogungbemi E, Ijaodola O, Khatib F, Durrant A et al (2019) A comprehensive study of the effect of bipolar plate (BP) geometry design on the performance of proton exchange membrane (PEM) fuel cells. Renew Sust Energ Rev 111:236–260
17. Nguyen TV (1996) A gas distributor design for proton-exchange-membrane fuel cells. J Electrochem Soc 143(5):L103
18. Tang A, Crisci L, Bonville L, Jankovic J (2021) An overview of bipolar plates in proton exchange membrane fuel cells. J Renew Sustain Energy 13(2):022701
19. Hüner B, Kıstı M, Uysal S, Uzgören IN, Ozdogan E, Süzen YO et al (2022) An overview of various additive manufacturing technologies and materials for electrochemical energy conversion applications. ACS Omega 7(45):40638–40658
20. Guo H, Chen L, Ismail SA, Jiang L, Guo S, Gu J et al (2022) Gas diffusion layer for proton exchange membrane fuel cells: a review. Materials 15(24):8800
21. Park S, Lee J-W, Popov BN (2006) Effect of carbon loading in microporous layer on PEM fuel cell performance. J Power Sources 163(1):357–363
22. Lettenmeier P, Kolb S, Sata N, Fallisch A, Zielke L, Thiele S et al (2017) Comprehensive investigation of novel pore-graded gas diffusion layers for high-performance and cost-effective proton exchange membrane electrolyzers. Energy Environ Sci 10(12):2521–2533
23. Mehta V, Cooper JS (2003) Review and analysis of PEM fuel cell design and manufacturing. J Power Sources 114(1):32–53
24. Zhang J, Shen S, Zhang J, Shen S (2021) Membrane electrode assembly (MEA). In: Low platinum fuel cell technologies. Springer, Berlin, pp 127–165
25. Bladergroen B, Su H, Pasupathi S, Linkov V (2012) Overview of membrane electrode assembly preparation methods for solid polymer electrolyte electrolyzer. Electrolysis 1:46–59
26. Kaya MF, Kıstı M, Hüner B, Demir N (2023) Electrochemical methods and materials for transition metal-based electrocatalysts in alkaline and acidic media. In: Transition metal-based electrocatalysts: applications in green hydrogen production and storage. ACS Publications, Washington, DC, pp 219–248
27. Bernt M, Siebel A, Gasteiger HA (2018) Analysis of voltage losses in PEM water electrolyzers with low platinum group metal loadings. J Electrochem Soc 165(5):F305–FF14
28. Zhong Y, Xia X, Shi F, Zhan J, Tu J, Fan HJ (2016) Transition metal carbides and nitrides in energy storage and conversion. Adv Sci 3(5):1500286
29. Anantharaj S, Ede SR, Sakthikumar K, Karthick K, Mishra S, Kundu S (2016) Recent trends and perspectives in electrochemical water splitting with an emphasis on sulfide, selenide, and phosphide catalysts of Fe, Co, and Ni: a review. ACS Catal 6(12):8069–8097
30. Li W, Gao X, Xiong D, Wei F, Song WG, Xu J et al (2017) Hydrothermal synthesis of monolithic Co3Se4 nanowire electrodes for oxygen evolution and overall water splitting with high efficiency and extraordinary catalytic stability. Adv Energy Mater 7(17):1602579
31. Li W, Wang X, Xiong D, Liu L (2016) Efficient and durable electrochemical hydrogen evolution using cocoon-like MoS2 with preferentially exposed edges. Int J Hydrog Energy 41(22):9344–9354
32. Wang Y, Zhao Y, Ding X, Qiao L (2021) Recent advances in the electrochemistry of layered post-transition metal chalcogenide nanomaterials for hydrogen evolution reaction. J Energy Chem 60:451–479
33. Amorim I, Xu J, Zhang N, Yu Z, Araujo A, Bento F et al (2021) Dual-phase CoP−CoTe2 nanowires as an efficient bifunctional electrocatalyst for bipolar membrane-assisted acid-alkaline water splitting. Chem Eng J 420:130454
34. Kim MJ, Cruz MA, Ye S, Gray AL, Smith GL, Lazarus N et al (2019) One-step electrodeposition of copper on conductive 3D printed objects. Addit Manuf 27:318–326
35. Li W, Xiong D, Gao X, Song W-G, Xia F, Liu L (2017) Self-supported Co-Ni-P ternary nanowire electrodes for highly efficient and stable electrocatalytic hydrogen evolution in acidic solution. Catal Today 287:122–129

36. Callejas JF, Read CG, Roske CW, Lewis NS, Schaak RE (2016) Synthesis, characterization, and properties of metal phosphide catalysts for the hydrogen-evolution reaction. Chem Mater 28(17):6017–6044
37. Gao J, Tao H, Liu B (2021) Progress of nonprecious-metal-based electrocatalysts for oxygen evolution in acidic media. Adv Mater 33(31):2003786
38. An L, Wei C, Lu M, Liu H, Chen Y, Scherer GG et al (2021) Recent development of oxygen evolution electrocatalysts in acidic environment. Adv Mater 33(20):2006328
39. Cherevko S, Geiger S, Kasian O, Kulyk N, Grote J-P, Savan A et al (2016) Oxygen and hydrogen evolution reactions on Ru, RuO2, Ir, and IrO2 thin film electrodes in acidic and alkaline electrolytes: a comparative study on activity and stability. Catal Today 262:170–180
40. Danilovic N, Subbaraman R, Chang K-C, Chang SH, Kang YJ, Snyder J et al (2014) Activity–stability trends for the oxygen evolution reaction on monometallic oxides in acidic environments. J Phys Chem Lett 5(14):2474–2478
41. Bertuccioli L, Chan A, Hart D, Lehner F, Madden B, Standen E (2014) Study on development of water electrolysis in the EU-final report. E4tech Fuel Cells & Hydrogen Joint Undertaking
42. Minke C, Suermann M, Bensmann B, Hanke-Rauschenbach R (2021) Is iridium demand a potential bottleneck in the realization of large-scale PEM water electrolysis? Int J Hydrog Energy 46(46):23581–23590
43. Shirvanian P, van Berkel F (2020) Novel components in proton exchange membrane (PEM) water electrolyzers (PEMWE): status, challenges and future needs. A mini review. Electrochem Commun 114:106704
44. Liu L (2021) Platinum group metal free nano-catalysts for proton exchange membrane water electrolysis. Curr Opin Chem Eng 34:100743
45. Shiva Kumar S, Ramakrishna S, Rama Devi B, Himabindu V (2018) Phosphorus-doped carbon nanoparticles supported palladium electrocatalyst for the hydrogen evolution reaction (HER) in PEM water electrolysis. Ionics 24:3113–3121
46. Feng Q, Liu G, Wei B, Zhang Z, Li H, Wang H (2017) A review of proton exchange membrane water electrolysis on degradation mechanisms and mitigation strategies. J Power Sources 366:33–55
47. Millet P (2016) Conventional and innovative electrocatalysts for PEM water electrolysis. ECS Trans 75(14):1073
48. Eftekhari A (2017) Electrocatalysts for hydrogen evolution reaction. Int J Hydrog Energy 42(16):11053–11077
49. Daiane Ferreira da Silva C, Claudel F, Martin V, Chattot R, Abbou S, Kumar K et al (2021) Oxygen evolution reaction activity and stability benchmarks for supported and unsupported IrO x electrocatalysts. ACS Catal 11(7):4107–4116
50. Anwar S, Khan F, Zhang Y, Djire A (2021) Recent development in electrocatalysts for hydrogen production through water electrolysis. Int J Hydrog Energy 46(63):32284–32317
51. Carmo M, Fritz DL, Mergel J, Stolten D (2013) A comprehensive review on PEM water electrolysis. Int J Hydrog Energy 38(12):4901–4934
52. Norazahar N, Khan F, Rahmani N, Ahmad A (2024) Degradation modelling and reliability analysis of PEM electrolyzer. Int J Hydrog Energy 50:842–856
53. Zaccarine SF, Shviro M, Weker JN, Dzara MJ, Foster J, Carmo M et al (2022) Multi-scale multi-technique characterization approach for analysis of pem electrolyzer catalyst layer degradation. J Electrochem Soc 169(6):064502
54. Khatib FN, Wilberforce T, Ijaodola O, Ogungbemi E, El-Hassan Z, Durrant A et al (2019) Material degradation of components in polymer electrolyte membrane (PEM) electrolytic cell and mitigation mechanisms: a review. Renew Sust Energ Rev 111:1–14
55. Ababao PM, Oh I (2023) Recent advances in catalyst materials for PEM water electrolysis. J Korean Electrochem Soc 26(2):19–34
56. Moschovi A, Zagoraiou E, Polyzou E, Yakoumis I (2021) Recycling of critical raw materials from hydrogen chemical storage stacks (PEMWE), membrane electrode assemblies (MEA) and electrocatalysts. IOP Conf Ser Mater Sci Eng.: IOP Publishing 1024:012008

57. Liu Q, Sun S, Zhang L, Luo Y, Yang Q, Dong K et al (2022) N, O-doped carbon foam as metal-free electrocatalyst for efficient hydrogen production from seawater. Nano Res 15(10): 8922–8927
58. Chen Z, Yun S, Wu L, Zhang J, Shi X, Wei W et al (2023) Waste-derived catalysts for water electrolysis: circular economy-driven sustainable green hydrogen energy. Nano-Micro Lett 15(1):4
59. Graś M, Kolanowski Ł, Chen Z, Lota K, Jurak K, Ryl J et al (2021) Partial inhibition of borohydride hydrolysis using porous activated carbon as an effective method to improve the electrocatalytic activity of the DBFC anode. Sustain Energy Fuels 5(17):4401–4413
60. Upadhyay S, Pandey O (2022) Low-temperature synthesized Mo2C and novel Mo2C–MnO2 heterostructure for highly efficient hydrogen evolution reaction and high-performance capacitors. J Power Sources 535:231450
61. Burton N, Padilla R, Rose A, Habibullah H (2021) Increasing the efficiency of hydrogen production from solar powered water electrolysis. Renew Sust Energ Rev 135:110255
62. Murugesan C, Senthilkumar B, Barpanda P (2022) Biowaste-derived highly porous N-doped carbon as a low-cost bifunctional electrocatalyst for hybrid sodium–air batteries. ACS Sustain Chem Eng 10(28):9077–9086
63. Deblonde GJ-P, Chagnes A, Cote G (2023) Recent advances in the chemistry of hydrometallurgical methods. Sep Purifi Rev 52(3):221–241
64. Chen Z, Wei W, Chen H, Ni B-J (2023) Eco-designed electrocatalysts for water splitting: a path toward carbon neutrality. Int J Hydrog Energy 48(16):6288–6307
65. Siracusano S, Hodnik N, Jovanovic P, Ruiz-Zepeda F, Šala M, Baglio V et al (2017) New insights into the stability of a high performance nanostructured catalyst for sustainable water electrolysis. Nano Energy 40:618–632
66. Peng J, Tao P, Song C, Shang W, Deng T, Wu J (2022) Structural evolution of Pt-based oxygen reduction reaction electrocatalysts. Chin J Catal 43(1):47–58
67. Aßmann P, Gago AS, Gazdzicki P, Friedrich KA, Wark M (2020) Toward developing accelerated stress tests for proton exchange membrane electrolyzers. Curr Opin Electrochem 21:225–233
68. Milosevic M, Böhm T, Körner A, Bierling M, Winkelmann L, Ehelebe K et al (2023) In search of lost iridium: quantification of anode catalyst layer dissolution in proton exchange membrane water electrolyzers. ACS Energy Lett 8:2682–2688
69. Burch M, Lewinski K, Buckett M, Luopa S, Sun F, Olson E et al (2021) A novel work-flow to study Ir electrode thinning and dissolution in proton exchange membrane water electrolyzers. J Power Sources 500:229978
70. Ouimet RJ, Glenn JR, De Porcellinis D, Motz AR, Carmo M, Ayers KE (2022) The role of Electrocatalysts in the development of gigawatt-scale PEM electrolyzers. ACS Catal 12(10): 6159–6171
71. Omeiza LA, Abdalla AM, Wei B, Dhanasekaran A, Subramanian Y, Afroze S et al (2023) Nanostructured electrocatalysts for advanced applications in fuel cells. Energies 16(4):1876
72. Indiarto R, Indriana LPA, Andoyo R, Subroto E, Nurhadi B (2022) Bottom–up nanoparticle synthesis: a review of techniques, polyphenol-based core materials, and their properties. Eur Food Res Technol 248(1):1–24
73. Joy J, Krishnamoorthy A, Tanna A, Kamathe V, Nagar R, Srinivasan S (2022) Recent developments on the synthesis of nanocomposite materials via ball milling approach for energy storage applications. Appl Sci 12(18):9312
74. Baig N, Kammakakam I, Falath W (2021) Nanomaterials: a review of synthesis methods, properties, recent progress, and challenges. Mater Adv 2(6):1821–1871
75. Miller JC, Haglund RF (1991) Laser ablation: mechanisms and applications. Springer, Berlin
76. Srivastava N, Srivastava M, Mishra P, Gupta VK (2020) Green synthesis of nanomaterials for bioenergy applications. Wiley, Hoboken
77. Purwidyantri A, Prabowo BA (2022) The prospects of colloidal lithography towards low-cost and scalable sensors. IEEE Nanotechnol Mag 16(5):20–28

78. Colson P, Henrist C, Cloots R (2013) Nanosphere lithography: a powerful method for the controlled manufacturing of nanomaterials. J Nanomater 2013:21
79. Biswas A, Bayer IS, Biris AS, Wang T, Dervishi E, Faupel F (2012) Advances in top–down and bottom–up surface nanofabrication: techniques, applications & future prospects. Adv Colloid Interf Sci 170(1–2):2–27
80. Gutiérrez-Cruz A, Ruiz-Hernández AR, Vega-Clemente JF, Luna-Gazcón DG, Campos-Delgado J (2022) A review of top-down and bottom-up synthesis methods for the production of graphene, graphene oxide and reduced graphene oxide. J Mater Sci 57(31):14543–14578
81. Harish V, Ansari MM, Tewari D, Gaur M, Yadav AB, García-Betancourt M-L et al (2022) Nanoparticle and nanostructure synthesis and controlled growth methods. Nanomaterials 12(18):3226
82. Banik A, Famprikis T, Ghidiu M, Ohno S, Kraft MA, Zeier WG (2021) On the underestimated influence of synthetic conditions in solid ionic conductors. Chem Sci 12(18):6238–6263
83. Parashar M, Shukla VK, Singh R (2020) Metal oxides nanoparticles via sol–gel method: a review on synthesis, characterization and applications. J Mater Sci Mater Electron 31:3729–3749
84. Wulandari I, Santjojo D, Shobirin R, Sabarudin A (2017) Characteristics and magnetic properties of chitosan-coated Fe3O4 nanoparticles prepared by ex-situ co-precipitation method. Rasayan J Chem 10:1348–1358
85. Trifoi AR, Matei E, Râpă M, Berbecaru A-C, Panaitescu C, Banu I et al (2023) Coprecipitation nanoarchitectonics for the synthesis of magnetite: a review of mechanism and characterization. React Kinet Mech Catal 136(6):2835–2874
86. Li J, Wu Q, Wu J (2016) 12: Synthesis of nanoparticles via solvothermal and hydrothermal methods. In: Handbook of nanoparticles. Springer, Cham
87. Gan YX, Jayatissa AH, Yu Z, Chen X, Li M (2020) Hydrothermal synthesis of nanomaterials. J Nanomater 2020:1–3
88. Liao X, Lu R, Xia L, Liu Q, Wang H, Zhao K et al (2022) Density functional theory for electrocatalysis. Energy Environ Mater 5(1):157–185
89. Chen Z, Chen L, Wen Z, Jiang Q (2019) Understanding electro-catalysis by using density functional theory. Phys Chem Chem Phys 21(43):23782–23802
90. Sargeant E, Illas F, Rodríguez P, Calle-Vallejo F (2022) On the shifting peak of volcano plots for oxygen reduction and evolution. Electrochim Acta 426:140799
91. He L, Yu G, Cheng Y, Wang N, Hu W (2023) Non-precious metal-based catalysts for water electrolysis to produce H 2 under industrial conditions. Mater Chem Front 7(22):5661–5692
92. Deng J, Deng D, Bao X (2017) Robust catalysis on 2D materials encapsulating metals: concept, application, and perspective. Adv Mater 29(43):1606967
93. Haslam GE, Chin X-Y, Burstein GT (2011) Passivity and electrocatalysis of nanostructured nickel encapsulated in carbon. Phys Chem Chem Phys 13(28):12968–12974
94. Chung HT, Won JH, Zelenay P (2013) Active and stable carbon nanotube/nanoparticle composite electrocatalyst for oxygen reduction. Nat Commun 4(1):1922
95. Hu K, Ohto T, Chen L, Han J, Wakisaka M, Nagata Y et al (2018) Graphene layer encapsulation of non-noble metal nanoparticles as acid-stable hydrogen evolution catalysts. ACS Energy Lett 3(7):1539–1544
96. Hüner B, Demir N, Kaya MF (2022) Electrodeposition of NiCu bimetal on 3D printed electrodes for hydrogen evolution reactions in alkaline media. Int J Hydrog Energy 47(24): 12136–12146
97. Zhang J, Wang G, Liao Z, Zhang P, Wang F, Zhuang X et al (2017) Iridium nanoparticles anchored on 3D graphite foam as a bifunctional electrocatalyst for excellent overall water splitting in acidic solution. Nano Energy 40:27–33
98. Thoi VS, Usiskin RE, Haile SM (2015) Platinum-decorated carbon nanotubes for hydrogen oxidation and proton reduction in solid acid electrochemical cells. Chem Sci 6(2):1570–1577

99. Xu J, Liu G, Li J, Wang X (2012) The electrocatalytic properties of an IrO2/SnO2 catalyst using SnO2 as a support and an assisting reagent for the oxygen evolution reaction. Electrochim Acta 59:105–112
100. Papazisi KM, Siokou A, Balomenou S, Tsiplakides D (2012) Preparation and characterization of IrxPt1−xO2 anode electrocatalysts for the oxygen evolution reaction. Int J Hydrog Energy 37(21):16642–16648
101. Nørskov JK, Bligaard T, Logadottir A, Kitchin J, Chen JG, Pandelov S et al (2005) Trends in the exchange current for hydrogen evolution. J Electrochem Soc 152(3):J23
102. Ramakrishna S, Reddy DS, Kumar SS, Himabindu V (2016) Nitrogen doped CNTs supported palladium electrocatalyst for hydrogen evolution reaction in PEM water electrolyser. Int J Hydrog Energy 41(45):20447–20454
103. Masson J, Molina R, Roth E, Gaussens G, Lemaire F (1982) Obtention and evaluation of polyethylene-based solid polymer electrolyte membranes for hydrogen production. Int J Hydrog Energy 7(2):167–171
104. Grigoriev S, Millet P, Volobuev S, Fateev V (2009) Optimization of porous current collectors for PEM water electrolysers. Int J Hydrog Energy 34(11):4968–4973
105. Vielstich W, Lamm A, Gasteiger H (2003) Handbook of fuel cells. Fundamentals, technology, applications. Wiley, Hoboken
106. Xu W, Liu C, Xing W, Lu T (2007) A novel hybrid based on carbon nanotubes and heteropolyanions as effective catalyst for hydrogen evolution. Electrochem Commun 9(1): 180–184
107. Du F, Zhang Y, He H, Li T, Wen G, Zhou Y et al (2019) Electrodeposited amorphous cobalt phosphosulfide on Ni foams for highly efficient overall water splitting. J Power Sources 431: 182–188
108. Giancola S, Zatoń M, Reyes-Carmona Á, Dupont M, Donnadio A, Cavaliere S et al (2019) Composite short side chain PFSA membranes for PEM water electrolysis. J Membr Sci 570: 69–76
109. Sun X, Simonsen SC, Norby T, Chatzitakis A (2019) Composite membranes for high temperature PEM fuel cells and electrolysers: a critical review. Membranes 9(7):83
110. Antonucci V, Di Blasi A, Baglio V, Ornelas R, Matteucci F, Ledesma-Garcia J et al (2008) High temperature operation of a composite membrane-based solid polymer electrolyte water electrolyser. Electrochim Acta 53(24):7350–7356
111. Ardizzone S, Bianchi C, Cappelletti G, Ionita M, Minguzzi A, Rondinini S et al (2006) Composite ternary SnO2–IrO2–Ta2O5 oxide electrocatalysts. J Electroanal Chem 589(1): 160–166
112. King LA, Hubert MA, Capuano C, Manco J, Danilovic N, Valle E et al (2019) A non-precious metal hydrogen catalyst in a commercial polymer electrolyte membrane electrolyser. Nat Nanotechnol 14(11):1071–1074
113. Giovanni CD, Reyes-Carmona Á, Coursier A, Nowak S, Grenèche JM, Lecoq H et al (2016) Low-cost nanostructured iron sulfide electrocatalysts for PEM water electrolysis. ACS Catal 6(4):2626–2631
114. Kelly TG, Hunt ST, Esposito DV, Chen JG (2013) Monolayer palladium supported on molybdenum and tungsten carbide substrates as low-cost hydrogen evolution reaction (HER) electrocatalysts. Int J Hydrog Energy 38(14):5638–5644
115. Bunea S, Urakawa A (2021) The influence of tin addition on the electrochemical performance of IrO2 and Ir black oxygen evolution catalysts in polymer electrolyte membrane electrolysis. In: Electrochemical Society meeting abstracts 239. The Electrochemical Society, Inc, Pennington, p 1194
116. Corona-Guinto J, Cardeño-García L, Martínez-Casillas D, Sandoval-Pineda JM, Tamayo-Meza P, Silva-Casarin R et al (2013) Performance of a PEM electrolyzer using RuIrCoOx electrocatalysts for the oxygen evolution electrode. Int J Hydrog Energy 38(28):12667–12673

117. Rangel-Martinez D, Nigam K, Ricardez-Sandoval LA (2021) Machine learning on sustainable energy: a review and outlook on renewable energy systems, catalysis, smart grid and energy storage. Chem Eng Res Des 174:414–441
118. Liu J, Zhang J, Sun H, Gao C, Chen Y, Cui R et al (2024) Ultrastable bifunctional multi-stage active metal catalysts for low concentration CO2 capture and in-situ conversion. Fuel 357: 129801
119. Chen Z, Guo L, Pan L, Yan T, He Z, Li Y et al (2022) Advances in oxygen evolution electrocatalysts for proton exchange membrane water electrolyzers. Adv Energy Mater 12(14): 2103670
120. Krishnan S, Corona B, Kramer GJ, Junginger M, Koning V (2024) Prospective LCA of alkaline and PEM electrolyser systems. Int J Hydrog Energy 55:26–41
121. Mori M, Iribarren D, Cren J, Cor E, Lotrič A, Gramc J et al (2023) Life cycle sustainability assessment of a proton exchange membrane fuel cell technology for ecodesign purposes. Int J Hydrog Energy 48:39673–39689
122. Standardization I (2006) Environmental management life cycle assessment principles and framework. ISO, Brussels, p 14040
123. Ciroth A (2021) Data quality in life cycle inventories. In: Life cycle inventory analysis: methods and data. Springer, Cham, pp 97–122
124. Mlakar N, Lotrič A, Sekavčnik M, Mori M (2022) The influence of degradation effects in proton exchange membrane fuel cells on life cycle assessment modelling and environmental impact indicators. Int J Hydrog Energy 47(57):24223–24241
125. Duclos L, Lupsea M, Mandil G, Svecova L, Thivel P-X, Laforest V (2017) Environmental assessment of proton exchange membrane fuel cell platinum catalyst recycling. J Clean Prod 142:2618–2628
126. Lozanovski A, Schuller O, Faltenbacher M, Bode M, Finkbeiner M, Chomkhamsri K (2011) Guidance document for performing LCAs on fuel cells and H_2 technologies. Retrieved February 10:2015
127. Van Allsburg KM, Tan EC, Super JD, Schaidle JA, Baddour FG (2022) Early-stage evaluation of catalyst manufacturing cost and environmental impact using CatCost. Nat Catal 5(4): 342–353
128. Rasul M, Hazrat M, Sattar M, Jahirul M, Shearer M (2022) The future of hydrogen: challenges on production, storage and applications. Energy Convers Manag 272:116326
129. Sadiq M, Alshehhi RJ, Urs RR, Mayyas AT (2023) Techno-economic analysis of Green-H2@ Scale production. Renew Energy 219:119362
130. Saba SM, Müller M, Robinius M, Stolten D (2018) The investment costs of electrolysis – a comparison of cost studies from the past 30 years. Int J Hydrog Energy 43(3):1209–1223
131. European Commission (2023) Repowereu: affordable, secure and sustainable energy for Europe. European Commission, Brussels
132. Government of United Kingdom (2023) Policy Paper British Energy Security Strategy
133. PlugPower: Clean Hydrogen Partnership to Promote Electrolyzers. https://www.plugpower.com/sk-es-plug-power-k-water-and-sk-plug-hyverse-established-clean-hydrogen-partnership-to-promote-electrolyzers/#:~:text=SK%20E%26S%20(CEO%20Choo%20Hyeongwook,a%20focus%20on%20promoting%20electrolyzers. (2023). Accessed 15/05/2024.
134. Zhou P, Niu P, Liu J, Zhang N, Bai H, Chen M et al (2022) Anodized steel: the most promising bifunctional electrocatalyst for alkaline water electrolysis in industry. Adv Funct Mater 32(26):2202068
135. Fund Green Climate: Strategic Plan for the Green Climate Fund 2024–2027. https://www.greenclimate.fund/document/strategic-plan-green-climate-fund-2024-2027. (2023). Accessed 15/04/2024.
136. IRENA: Green hydrogen for sustainable industrial development. https://www.irena.org/Events/2024/Mar/IRENA-Policy-Talks-2024-Green-hydrogen-for-sustainable-industrial-development. (2024). Accessed 15/05/2024.

137. Central Hydrogen: Breaking the Finance Barrier for Hydrogen and Carbon Capture. https://hydrogen-central.com/bcg-breaking-finance-barrier-hydrogen-carbon-capture/. (2023). Accessed 15/05/2024.
138. JChemanalyst: Japan Ambitious $107 Billion Plan to Harness Hydrogen: A Green Revolution in Sight. https://www.chemanalyst.com/NewsAndDeals/NewsDetails/japan-ambitious-107-billion-plan-to-harness-hydrogen-a-green-revolution-in-sight-17585. (2023). Accessed 15/03/2024.
139. Taghizadeh-Hesary F, Li Y, Rasoulinezhad E, Mortha A, Long Y, Lan Y, et al. Green finance and the economic feasibility of hydrogen projects. International Journal of Hydrogen Energy. 2022;47(58):24511–24522.
140. Taie Z, Peng X, Kulkarni D, Zenyuk IV, Weber AZ, Hagen C et al (2020) Pathway to complete energy sector decarbonization with available iridium resources using ultralow loaded water electrolyzers. ACS Appl Mater Interfaces 12(47):52701–52712
141. Islam J, Yoon BS, Thien PT, Ko CH, Kim S-K (2024) Enhancing oxygen evolution reaction performance in PEM water electrolyzers through support pore structure optimization. Catal Today 425:114349
142. Wang C, Feng L (2024) Recent advances and perspectives of Ir-based anode catalysts in PEM water electrolysis. Energy Adv 3:14

The Concept of Electrochemical Hydrogen Compression and Purification Technology

Doria Marciuš

Acronyms

ATR	Autothermal reformer
BEV	Battery electric vehicle
BOP	Balance of plant
CAPEX	Capital expenditure
CGH2	Compressed gaseous hydrogen
CL	Catalyst layer
EHC	Electrochemical hydrogen compressor
EHSP	European Hydrogen Safety Panel
FCEV	Fuel cell electric vehicles
FMEA	Failure mode and effect analysis
FTA	Fault tree analysis
GDL	Gas diffusion layer
H2ME	Hydrogen mobility Europe
HAZOP	Hazard operability analysis
HER	Hydrogen evolution reaction
HHV	Higher heating value
HOR	Hydrogen oxidation reaction
HRS	Hydrogen refueling station
HySEA	Hydrogen safety for energy applications
LH2	Liquid hydrogen
LOHC	Liquid organic hydrogen carrier
MEA	Membrane electrode assembly

D. Marciuš (✉)
Faculty of Mechanical Engineering and Naval Architecture, University of Zagreb, Zagreb, Croatia
e-mail: doria.marcius@fsb.unizg.hr

A. Kovač (ed.), *Challenges and Solutions in the Hydrogen Value Chain*,
https://doi.org/10.1007/978-3-031-69051-8_4

MTBF	Mean time between failures
NFPA	National Fire Protection Association
OPEX	Operating and maintenance expenditure
PEM	Proton exchange membrane
PEMEL	Proton exchange membrane electrolyzer
PEMFC	Proton exchange membrane fuel cell
PRD	Pressure release device
PSA	Pressure swing adsorption
PSEHS	Pressure swing electrochemical hydrogen separation
RES	Renewable energy sources
RWGS	Reverse water gas shift
SAE	Society of Automotive Engineers International
SOFC	Solid oxide fuel cells
TCO	Total cost of ownership
TSA	Temperature swing adsorption
UV	Ultraviolet
VSA	Vacuum swing adsorption

Abbreviations

C	Carbon
CH_4	Methane
CO	Carbon monoxide
CO_2	Carbon dioxide
CuNi	Copper-nickel
e^-	Electron
H^+	Hydrogen proton
H_2	Hydrogen
H_2O	Water
H_2S	Hydrogen sulphide
H_{ad}	Hydrogen adsorbed-protons
He	Helium
N_2	Nitrogen
NH_3	Ammonia
Pt	Platinum
Ti	Titanium

1 Background of Hydrogen Compression

The background discussion elaborates on the electrochemical hydrogen compression research requirement and puts it into perspective based on the present most commonly used compression technology. It encompasses hydrogen properties relevant to compression, discusses the advantages of hydrogen compression compared to other hydrogen storage solutions, and describes different hydrogen compression methods in regard to electrochemical hydrogen compression.

1.1 Hydrogen Compression Potential

Hydrogen's growing influence in the energy market, as well as its increased and diversified usage, instigated research focus on topics such as hydrogen storage and transfer due to hydrogen's low density. For example, if hydrogen is used in fuel cell electric vehicles (FCEVs), its low density necessitates a large volume of stored hydrogen in a vehicle to enable an adequate driving range. The specific volume (the inverse of density expressed as the amount of volume per unit mass) of hydrogen gas amounts to 11.9 m^3/kg at 20 °C and 1 atm, while the specific volume of liquid hydrogen is 0.014 m^3/kg at −253 °C and 1 atm. Another way of expressing hydrogen relative density is via specific gravity, which is the ratio of the density at the same temperature and pressure of hydrogen to that of a reference substance (water for liquids with a density of 1000 kg/m^3, and air for gases with a density of 1.203 kg/m^3), which therefore has a specific gravity of 1.0 relative to itself. Hydrogen thus has a specific gravity of 0.0696 in a gaseous state, which is approximately 7% the density of air, and 0.0708 in a liquid state, which is also approximately 7% the density of water. In addition to specified hydrogen properties, another property relevant to compression is the hydrogen liquid-to-gas expansion ratio, which amounts to 1:848, meaning that hydrogen in its gaseous state at atmospheric conditions occupies 848 times more volume than it does in its liquid state. When compressed to 250 barg at atmospheric temperature, its expansion ratio to atmospheric pressure is 1:240. Hence, a higher storage pressure increases the expansion ratio, but hydrogen in a gaseous state under no conditions can be equal to the expansion ratio of liquid hydrogen [1]. When comparing the net calorific value at standard pressure and temperature conditions of conventional fuels such as methane, propane, gasoline, diesel, and methanol, hydrogen has the highest mass-energy density. Table 1 shows compared values of lower heating value (LHV) energy densities of hydrogen and its comparative fuels.

For example, 1 kg of hydrogen has 120 MJ/kg of energy, while 1 kg of gasoline has 43.4 MJ/kg, and 1 kg of natural gas has 47.1 MJ/kg. However, gaseous hydrogen

Table 1 Volumetric energy density of various fuels

Fuel	Volumetric energy density (kWh/m^3)
Hydrogen (gas at 1 atm and 15 °C)	2.79
Methane (gas at 1 atm and 15 °C)	9.04
Propane (gas at 1 atm and 15 °C)	24.08
Methanol (liquid)	4389
Gasoline (liquid)	8653
Diesel (liquid)	8732

Data from [1]

in a 460 L tank at 350 bar or hydrogen in a 340 L tank at 700 bar is an energy equivalent to gasoline in a 50 L tank. Hence, hydrogen's low volumetric energy density (0.01079 MJ/L at 1 bar and 15 °C) in the gaseous state makes it difficult to store and transfer, as for the same amount of energy, larger volumes of hydrogen need to be moved [2].

1.2 Hydrogen Storage Methods

As stated, due to its significantly low volumetric energy density, hydrogen has to be treated in some form to economically reduce its volume when being transported or stationary stored. The present most used storage options are compressed gaseous hydrogen (CGH2), liquid hydrogen (LH2)—liquefaction, the use of a liquid organic hydrogen carrier (LOHC), and conversion into ammonia, methanol, or synthetic fuels [3]. The theoretical volumetric density of LH2 amounts to 70 g/L at −253 °C and 1 atm, while for compressed gaseous hydrogen at room temperature, it amounts to 24 and 40 g/L at 350 and 700 bar, respectively. However, cooling hydrogen to such low temperatures requires considerable amounts of energy, which consume 25–40% of the hydrogen energy content, while compressing hydrogen requires 10% [4]. Recent hydrogen storage research has focused on a process of cryo-compression to store hydrogen at cryogenic temperatures in a vessel that can be pressurized. The idea is to combine the attributes of compressed and cryogenic hydrogen storage without the disadvantages of compressed hydrogen storage, which are the large volumes and high pressures required, and the disadvantages of LH2, which are the boil-off losses (up to 0.4% per day [4]). In this regard, the volumetric hydrogen storage capacity is increased from 70 g/L at 1 bar to 87 g/L at 240 bar by compressing liquefied hydrogen at 20 K [5]. Despite being a promising storage solution, the availability and cost of needed infrastructure are the main disadvantages, limiting its usage. Each of these mentioned hydrogen storage methods increases hydrogen's energy density by volume, as seen in Table 2.

Hydrogen storage vessels should encompass high volumetric and gravimetric energy densities, operation at room temperature and atmospheric pressure so that no further energy expenses are needed, fast uptake and release of hydrogen, safe and reliable usage without leakage, and balanced cost-effectiveness. All described hydrogen storage technologies have significant drawbacks, including complex, energy-intensive thermal management systems, gradual boil-offs or hydrogen embrittlement problems, poor efficiency and energy densities, expensive catalysts,

Table 2 Volumetric energy density of various hydrogen storage options

Hydrogen storage methods	CGH2 (15 °C, 300 bar)	CGH2 (15 °C, 500 bar)	CGH2 (15 °C, 700 bar)	LOHC	LH2	Ammonia	Methanol
Volumetric energy density (kWh/m^3)	733	1000	1333	1400	2367	3580	4378

Data from [3]

high investment costs, stability issues, slow response rates, high operating pressures, and dangerous risks of uncontrolled hazards. Regardless of its limitations, the current industry standard of compressed hydrogen storage offers the most functional solution to other technologies [6]. To sum up, compressing hydrogen is the most well-established and widely used hydrogen storage process, as the liquefaction of hydrogen brings about high cost, boil-off as a major disadvantage, and poor conversion efficiency. Compressed hydrogen is stored in pressure vessels that are classified according to types (Type I, Type II, Type III, and Type IV) differentiated based on material, pressure, cost, and gravimetric density. The most used materials for vessel construction are carbon fiber and nylon-6 [6].

1.3 Hydrogen Compression and Purification Methods

1.3.1 Hydrogen Compression

Reviewing hydrogen storage methods leaves the focus on hydrogen gas compression. Classified into mechanical and non-mechanical compressions, several methods are used for hydrogen compression, and the choice of method depends on factors such as the pressure levels, flow rates, efficiency, and the specific needs of the application. Mechanical compressors are the most common and can be used in various configurations, such as single-stage or multi-stage compression, depending on the required discharge pressure. They compress hydrogen by directly using mechanical energy to reach high pressures while achieving relatively high flow rates. The non-mechanical category consists of compressors with a limited or no number of moving parts that achieve compression mainly through electrochemical reactions. The ongoing research and development in the field of hydrogen technologies led to advancements in compressor technology, including improvements in efficiency and reliability, and the development of new compression methods, such as non-mechanical compressors. Table 3 provides short descriptions and applications of different hydrogen compression methods. State of the art of hydrogen compression technologies includes mechanical compressors, i.e., reciprocating, diaphragm, rotary screw, centrifugal, hydraulic piston, linear, and ionic liquid compressors, as well as innovative non-mechanical hydrogen technologies, such as cryogenic, metal hydride, electrochemical, and adsorption compressors [7].

The choice of compression method depends on various factors, including the required pressure level, energy efficiency, maintenance requirements, and the specific application in which the compressed hydrogen will be used. Each method has its advantages and limitations, and the selection is often based on a trade-off between factors like capital cost, operational cost, and efficiency. Presently, the most used compressors in hydrogen systems are generally either reciprocating piston or diaphragm compressors, both of which are conventional mechanical systems. Although reciprocating compressors present a mature solution, the extensive use of moving parts is the reason for increased manufacturing of complex designs, high maintenance cost due to the higher frequency of activities, and higher noise and vibration.

Table 3 Overview of different hydrogen compression methods (mechanical and non-mechanical)

Mechanical compressors	
Reciprocating piston compressors:	These compressors use a piston-cylinder arrangement to compress hydrogen gas. They are widely used in industrial applications and are suitable for both small and large-scale compression. The most typical problems with reciprocating piston compressors are sealing and lubrication-related issues. Lubrication with oil provides a risk of hydrogen impurity, as well as leakage. However, seal ring failures are the main limitation of high-speed oil-free reciprocating compressors. Also, the compressor's moving parts result in a shorter life, and higher operation and maintenance costs [8]
Rotary screw compressors:	Rotary screw compressors are a type of positive displacement compressors that use rotating helical screws to compress hydrogen gas. They are often used for medium-capacity compression and are known for their reliability and efficiency. The drawback is the need for an air-cooled or water-cooled heat exchanger to cool the compressed hydrogen gas [9]
Centrifugal compressors:	Centrifugal compressors use a rotating impeller to accelerate hydrogen gas, and then a diffuser slows the gas down, converting kinetic energy into pressure. They are commonly used in large-scale industrial applications but due to hydrogen's low molecular weight, they require high impeller tip speeds, which, therefore requires the use of high-strength materials [10]
Hydraulic piston compressors:	In hydraulic piston compressors, a piston is driven by hydraulic fluid to compress hydrogen gas. This method is often used for high-pressure compression. The advantage of a free piston is the reduced number of moving parts, but the sealing, complex design, and lubrication remain the issue for compressing hydrogen [8]
Diaphragm compressors:	Diaphragm compressors use a flexible diaphragm or membrane to separate the hydrogen gas from the compression mechanism. The diaphragm moves back and forth, alternately compressing and expanding the gas chamber. They are often used in applications where contamination or oil-free compression is critical. They are suitable for low to moderate pressure ranges [8]. However, the diaphragm is prone to fatigue fracture, resulting in the compressor's lifetime of a maximum of 6000 h [11]
Linear compressors:	Linear compressors, also known as free-piston compressors, operate without traditional pistons. Instead, a linear motor or electromagnetic drive moves a compression element back and forth to compress the hydrogen gas directly. They are known for their efficiency and are used in various applications, including hydrogen fueling stations and small-scale compression systems. The initial cost of linear compressors can be higher than that of rotary compressors and can produce more vibration and noise during operation due to the necessity to operate at resonant conditions [7]
Ionic liquid compressors:	Ionic liquid hydrogen compressors use ionic liquids, which are salts in a liquid state at relatively low temperatures, as the working fluid. These compressors typically operate at lower temperatures compared to traditional compressors. They are being explored as an alternative to conventional methods as they replace the cranked connecting rod piston with a liquid column and a free piston

(continued)

Table 3 (continued)

	[8]. They have the potential for energy-efficient compression, and research regarding efficiency increase includes using porous media and internal liquid-spray cooling to optimize their performance [11]
Non-mechanical compressors	
Absorption/adsorption compressors:	Absorption and adsorption processes involve using a solid material (absorbent or adsorbent) to selectively take up hydrogen gas. The absorbed or adsorbed hydrogen can be released by changing conditions such as temperature or pressure. This method is often used for onboard hydrogen storage in fuel cell vehicles. The disadvantages of adsorption compressors are the low thermal conductivity of adsorbents, difficulty in thermal management, and required low-temperature operation [12]
Membrane compressors:	Membrane compressors use semi-permeable membranes to selectively allow hydrogen to pass through, separating it from other gases and increasing its pressure. This method is suitable for moderate pressure applications, but not higher pressures [12]
Cryogenic compressors:	Cryogenic hydrogen compression involves cooling hydrogen gas to very low temperatures, typically below its boiling point to convert it into a liquid state, after which hydrogen is pumped to achieve higher pressures. Cryogenic compression is often used when extremely high pressures are required, such as in rocket propulsion systems [13]. The process is energy-intensive due to the need for cryogenic refrigeration and it is difficult to manage thermal insulation [14]
Metal hydride compressors:	Metal hydride compressors use the ability of certain metals to form hydrides with hydrogen under suitable conditions. The metal hydride absorbs hydrogen gas at low pressures and releases it at higher pressures when the temperature is raised, or other conditions are met. Metal hydride compressors offer advantages such as reversible hydrogen storage and can operate at relatively mild temperatures and pressures. On the other hand, they are relatively immature technology in terms of market readiness, offer relatively low efficiency, and are heavy [15]
Electrochemical hydrogen compressor:	Electrochemical hydrogen compressors (EHCs) utilize electrochemical reactions (the hydrogen oxidation at the anode and its reduction at the cathode) for hydrogen compression in a closed volume. The technology is based on the same principles as those of a proton exchange membrane fuel cell (PEMFC)

Furthermore, reciprocating compressors are not efficient at high flow rates as mechanical stresses limit the use of higher speeds, restricting the flow rates. Diaphragm compressor's main disadvantage is a potential diaphragm failure due to radial stress related to diaphragm deflection [10]. Among mechanical compressors, liquid compressors are generally considered to be the most suitable for ultra-high-pressure compressions as their performance during the compression process is close to that of the isothermal process [11]. To sum up, mechanical compressors are more suitable for high-power applications nominally above 100 kW but are not as efficient as non-mechanical compressors for low-power applications. Furthermore, mechanical compressors' limitations often include corrosion, hydrogen impurity, low theoretical efficiency, and frequent and high-cost maintenance due to moving parts that cause mechanical friction, noise, and vibrations [16]. For this reason, non-mechanical compressors are presently the focus of ongoing research in hydrogen compression studies. Out of all the non-mechanical compression technologies described, electrochemical hydrogen compressors (EHCs) stand out due to having various advantages, such as isothermal process, noiseless and vibration-free operation, modularity to combine cells in a stack with either high hydrogen pressure or high flow rate, non-existence of moving parts, and high efficiency (usually over 60%), but since the technology is not yet mature, much progress in development and research is still needed [2].

1.3.2 Hydrogen Purification

In addition to compression, for hydrogen to become a widespread energy vector for renewable energy sources (RES), its purification is also an unavoidable, energy-intensive process. Hydrogen purification technologies are methods used to separate, purify, and refine hydrogen gas from various sources to meet specific quality and purity requirements. The purity of hydrogen is crucial for its use in various applications, including fuel cells, industrial processes, and electronics. Presently, several technological approaches are used to extract hydrogen from gas mixtures, utilizing various characteristics of hydrogen under different industrial conditions. Common approaches for hydrogen recovery are based on adsorbing and absorbing the impurities (e.g., pressure swing adsorption (PSA), vacuum swing adsorption (VSA), and temperature swing adsorption (TSA)), condensing the impurities (e.g., cryogenic distillation), unconventional methods (e.g., gas centrifuges, and selective oxidation), and selectively permeable membranes (e.g., EHC—polymeric membrane, different types of membrane). Table 4 shows some common hydrogen purification technologies.

PSA and membrane separation processes are the most common commercial processes for hydrogen separation, with PSA being the most extensively used state-of-the-art industrial process (approx. 85% of the produced hydrogen is purified by PSA). PSA is most frequently used in the chemical/petrochemical industry, as well as to recover hydrogen from industrial-rich exhaust gases, including reforming off-gases, coke oven gases, and pyrolysis effluent gases [22]. Pressure-driven membranes are considered the most suitable option for hydrogen production because they are not as energy-intensive, and they yield high-purity hydrogen. PSA and cryogenic distillation technologies require multiple units and, in some instances, may involve supplementary wash columns for the removal of CO and CO_2 [23]. The cryogenic distillation process is widely used in many industries for high degree purification of industrial gases, including oxygen, nitrogen, helium, argon, natural gas, and hydrogen. This method is attractive due to the higher hydrogen recovery that can be achieved at moderate hydrogen purity yields ($\leq$95%). Similar to PSA, cryogenic distillation is perfect for large industrial scales but unsuitable for small portable

Table 4 Overview of different hydrogen purification methods

Pressure swing adsorption (PSA):	PSA is a cyclic adsorption process based on the ability of certain materials, such as activated carbon or zeolites, to selectively adsorb gases under pressure. Hydrogen is passed through an adsorbent bed at high pressure, where impurities are adsorbed, while hydrogen molecules are allowed to pass through. The pressure is then reduced, and the adsorbent is regenerated, releasing the impurities [17]
Membrane separation:	Membrane separation involves the use of selectively permeable membranes that allow the passage of hydrogen while blocking impurities based on differences in molecular size and affinity. Hydrogen gas is pushed through the membrane, and impurities are retained. This method is often used for removing contaminants like carbon dioxide and methane [18]
Cryogenic distillation:	Cryogenic distillation takes advantage of the different boiling points of gases at low temperatures. Hydrogen is cooled to cryogenic temperatures, and impurities are separated based on their boiling points. The hydrogen gas is liquefied and then distilled to separate impurities such as nitrogen, oxygen, and argon. The purified hydrogen is then vaporized and collected [19]
Polymeric membrane separation:	Similar to traditional membrane separation, polymeric membranes selectively allow hydrogen molecules to permeate through while blocking impurities. The membranes in this case are made from specialized polymers that exhibit high selectivity for hydrogen, making them suitable for applications where high purity is required [20]
Electrochemical hydrogen compressor (EHC):	Purification via EHC is an electrochemical process that selectively transports hydrogen ions through an electrolyte membrane. Under the influence of an electric potential, hydrogen ions move through the membrane, while impurities are blocked. The separated hydrogen is collected on the other side
Selective oxidation:	This method involves selectively oxidizing impurities present in the hydrogen stream. Common impurities such as carbon monoxide (CO) can be selectively oxidized to carbon dioxide (CO_2), which can then be removed using other purification methods [21]

applications [18]. Types of membranes for hydrogen separation are organic (polymeric), inorganic, and mixed matrix (hybrid). Membranes can be evaluated based on permeance (flux or permeation rate) and selectivity. The higher the selectivity, the lower the driving force required to achieve a certain separation—thereby reducing the cost of operation of the system. On the other hand, the higher the permeance, the active membrane area is the smaller, which also reduces the cost [24]. Advantages of membrane technologies include low energy consumption, the possibility of continuous operation, cost-effectiveness, low maintenance, and scalability [25]. However, despite their numerous advantages, membrane systems commonly depend on high-pressure feed streams, hydrogen embrittlement is often experienced, and the output hydrogen is located in the permeate (low-pressure stream after membrane), which requires additional compression in order for hydrogen to be used for transport and storage purposes. The additional process of compression brings forth high energy expenses and necessary additional equipment. Hydrogen purity achieved after the purification process via PSA ranges from 98 to 99.999 vol.%, membrane technology is capable of yielding hydrogen with a purity of >98 vol.%, and hydrogen purity after cryogenic distillation is in the range of 95–99.8 vol.%. Hydrogen recovery for the same purification technologies amounts to up to 90%, up to 99%, and up to 98%, respectively, while minimum feed purity required is >40%, >25%, and from 15 to 80%, respectively, for PSA, membrane, and cryogenic distillation [26].

Purification via EHCs is based on proton exchange, and polymeric membranes (PEMs) are seen as a promising alternative to pressure-driven membranes. When an inlet stream contains a gas-hydrogen mixture, hydrogen is selectively transferred through the proton-conducting electrolyte with high selectivity and low permeability in the form of protons, forming the high-purity hydrogen (up to 99.999 vol.%) for the outlet stream in a one-step operation. EHCs have multiple advantages over other conventional separation systems. The hydrogen separation rate can be controlled by the current due to Faraday's Law, and high hydrogen separation with high efficiency can be achieved at low cell voltages. Furthermore, as already mentioned, EHC acts simultaneously as a hydrogen purification and compression system with a high hydrogen collection rate. An additional important benefit of purification via EHC is that CO_2 can be concentrated, captured, and stored without any further treatment, which enables new possible applications regarding greenhouse gas emissions reductions [27]. The choice of hydrogen purification technology depends on factors such as the source of hydrogen, required purity levels, and the specific application. Often, a combination of these technologies is used in a purification system to achieve the desired hydrogen quality. Out of all described hydrogen separation technologies, EHCs stand out due to high selectivity, sensitivity to catalyst deactivation (e.g., CO deactivation), higher tolerance to "active" impurities (e.g., CO and CO_2) at higher temperatures, the hydrogen flux that can be controlled by the current, and simultaneous hydrogen separation and compression that can be achieved. However, this technology is still relatively new and should be researched more [26].

2 Electrochemical Hydrogen Compressor Technology Layout

This chapter provides the EHC technology layout that will focus on EHC design and operating principle, providing valuable insight into the technology, starting with its brief history and statistics of coming onto the market. Research regarding EHC will be discussed, as well as the fundamental principles EHC is based on and possible configurations of combining EHC cells into a stack—its advantages and drawbacks.

2.1 The Overview of Technology Rise

There have been scientific publications on the subject of hydrogen pumping using ion exchange membranes since the 1980s. However, there was no significant profit seen in it or other economic interest at the time until the development and progress in fuel cell and electrolyzer technology. In order to bypass expensive subsequent mechanical hydrogen compression, there have been attempts to modify proton exchange membrane electrolyzers (PEMEs) with thicker membranes to withstand high pressures. The incorporation of such membranes would allow PEME to simultaneously produce hydrogen and compress it. However, when implementing a thicker PEM, the Nernst potential at high pressures reaches the same order of magnitude as ohmic overpotential, thus resulting in a substantial electrolyzer efficiency decrease. On the other hand, when thinner membranes with support for higher pressures are used, lower resistances are detected, but with potential hazardous risks that present a critical safety issue. That is, when the pressure is increased, thin membranes consequently allow the permeation of hydrogen molecules through the membrane to the oxygen side, and vice versa, resulting in a hydrogen/oxygen mix. This gas mixture can potentially be explosive if a lower flammability limit is exceeded. Moreover, when a thinner membrane is used, the crossover flow causes notable losses in hydrogen production. To conclude, neither thick nor thin membranes present a solution for hydrogen compression via PEME, and therefore there is a requirement for other ways of electrochemical hydrogen compression—EHC technology. As seen in Fig. 1, the number of published articles regarding EHC per year reached 120 by the end of 2021, and the upcoming research trend is expected to

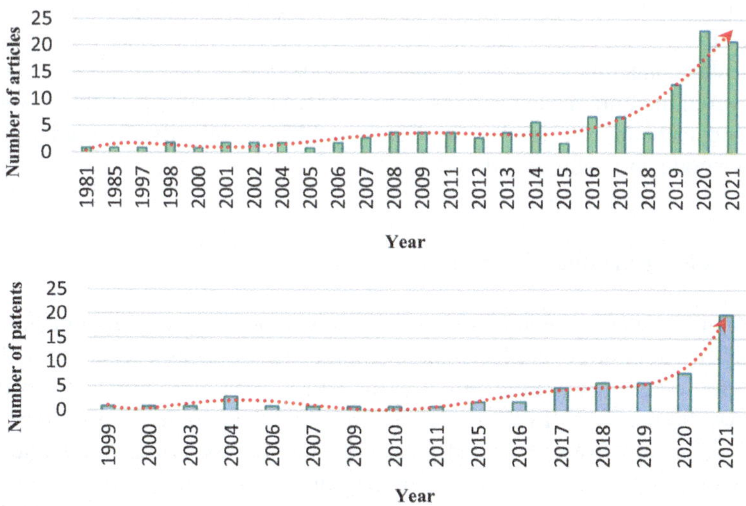

Fig. 1 History of published EHC scientific articles (≤2021) [2]

increase even more in the following years due to hydrogen's rising role in the energy transition to green energy.

The number of patents related to the EHC technology in cell or stack configuration reached 60 by the end of 2021, and the published scientific research also follows an upward trend. The most prominent uprise has been evident in the last few years, presumably due to the hydrogen requirement to be compressed to pressures up to 700 bar, which is needed for the fast-developing market of PEMFC technology and its implementation in FCEVs. Consequently, the growth of hydrogen road maps necessitates the expansion of hydrogen refueling station (HRS) infrastructure. As discussed in Sect. 3, presently, the most used compressors in hydrogen systems are generally either reciprocating piston or diaphragm compressors, both of which are conventional mechanical systems. These mechanical compressors account for more than 50% of the CAPEX (capital expenditure) in a hydrogen refueling station [27]. Hence, the pressure for researchers to develop more beneficial and profitable hydrogen compression technologies has started to rise, further pushed EHC onto the hydrogen market. Moreover, the standards for FCEV's hydrogen purity have heightened, which has brought EHC technology more favor as it has the potential to meet those standards alongside hydrogen compression. Currently, the EHC's importance is growing with the process of decarbonization, as hydrogen is recognized as an invaluable energy vector for renewable energy sources.

2.2 *Cell Components and Design*

The technology of EHC is designed on similar principles to the PEMFC. However, EHC has a dual function—hydrogen compression and purification—due to the selectivity of PEM (more will be explained in Sect. 4.3). As presented in Fig. 2, the main EHC components starting from the outside are end plates, electric insulators, current collectors, flow field plates, gaskets, gas diffusion layers (GDL), and three-layered membrane electrode assembly (MEA). Three-layered MEA consists of anode and cathode catalyst layers (CL) and PEM, while five-layered MEA also includes anode and cathode side GDLs.

2.2.1 Proton Exchange Membrane

The most integral component, a selective hydrogen proton conductive material at the center of the technology, is PEM. The membrane performs as a separator and a solid electrolyte. The most frequently used PEM type for EHC is Nafion—a Perfluorosulfonic acid (PFSA) material for low-temperature operations, whose thickness can vary from 51 to 254 μm. Generally, the experiments conducted with ECH contain hydrogen at room temperature, or its maximum temperature is limited to approx. 80 °C due to the highest decrease in membrane moisture level with high temperatures, which strongly reflects on the membrane proton conductivity.

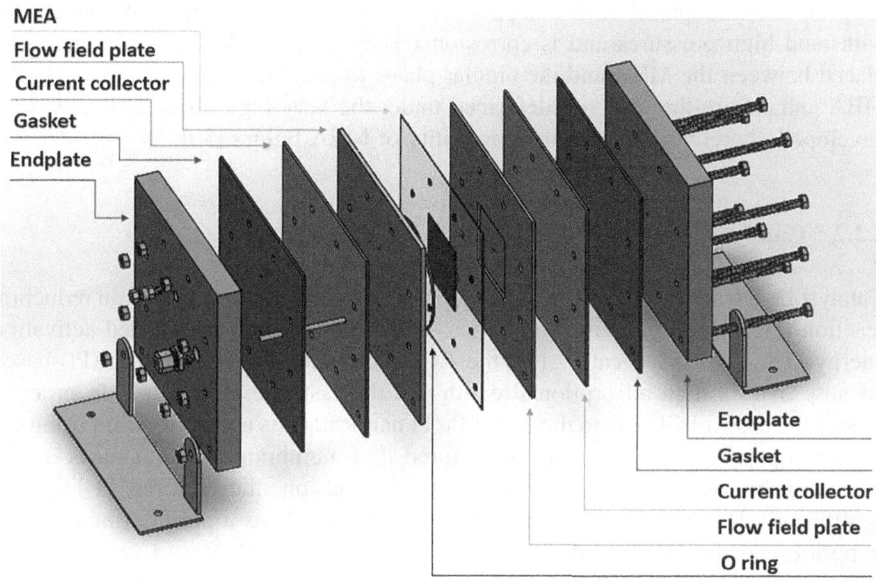

Fig. 2 Single-cell EHC components

Furthermore, one should keep in mind that at current densities above 0.5 A/cm^2, PEM requires an even greater amount of water due to heightened resistances that generate heat. A study showed that while testing EHC at 80 °C, there was an increase in the voltage necessary to maintain the applied current density, compared to the tests at 30 °C. Furthermore, at the same temperatures, EHC with a higher relative humidity (100% RH) than PEM showed better performance [28]. Another factor to consider regarding optimal PEM conditions is hydrogen molecule back diffusion. Hydrogen molecules transfer from the cathodic to the anodic compartment due to high pressure forming at the cathode side, thus limiting the maximum hydrogen output pressure. The most relevant factor to consider for the rate of hydrogen back diffusion (except for the cathode pressure) is temperature, after which relative humidity plays the biggest role. By raising the temperature, the hydrogen back diffusion rate significantly increases due to the increase in the solubility and the diffusion coefficient of hydrogen in the membrane. Also, by raising the relative humidity of PEM, the number of water molecules contained in the membrane increases, which modifies the intermolecular distance in the polymer chain and creates additional transport routes for hydrogen molecules to transfer from the cathode to the anode compartment [29]. To sum up, the ideal properties of PEM for EHC are high conductivity for hydrogen protons and low conductivity for impurities, high mechanical strength due to high-pressure differences between the anode and cathode sides, and low gas permeability to decrease the hydrogen back diffusion. One of the solutions for preventing PEM deformation, as well as MEA deformation, is incorporating additional materials of high strength and durability as support. An example of such a support structure is titanium (Ti) mesh, which can withstand high pressures and is corrosion resistant. A stainless-steel mesh can be placed between the MEA and the bipolar plates to provide structural support to the MEA and ensure that it is not deformed under the very high differential pressures developed, thereby mitigating the possibility of MEA failure [30].

2.2.2 Catalyst Layers

Catalyst layers enhance hydrogen oxidation reaction (HOR) and hydrogen reduction reaction (HER) electrochemical kinetics, which greatly reduces needed activation energy. The most used catalyst in the EHC design of CL is platinum (Pt)-based because of Pt's high adsorption strength and the fast rate of reaction. In order to reduce the costs of CL due to the use of Pt, Pt nanocatalysts are synthesized using the atomic deposition method or nanostructured thin membranes, with a focus on the even distribution of catalyst particles. For this reason, the required Pt load has significantly lowered over the years. Due to EHC's role as a purifier, the effect of impurities on the HOR in the anode should be monitored. E.g., CO_2 negatively affects the Pt catalyst. It is noted that an electrochemical reduction of CO_2, responsible for CO formation, may occur on the catalyst, hindering the hydrogen splitting and subsequent permeation through the membrane [31]. Thus, a catalyst on the anode side that operates with impurities alongside hydrogen must demonstrate very high HOR activity and be tolerant to the poison molecules. To sum up, the overall role of the CLs is to provide a surface for the electrochemical reactions to occur efficiently and to facilitate the transport of protons from the anode to the cathode through the PEM. The design and optimization of catalyst layers are critical for achieving high performance, efficiency, and durability in EHCs.

2.2.3 Gas Diffusion Layers

Gas diffusion layers are porous structures made of conductive materials with a thickness in the range of 100–400 μm located between the catalyst and the flow field. GDLs provide support to the catalyst nanoparticles as they can aggregate during the electrochemical reaction and enforce uniform distribution of reactants to the active area. The most used GDL materials are carbon-based (carbon paper or cloth) and consist of macroporous and microporous layers, but can also be metallic, such as Ti frit. For example, double GDLs can be used on both the anode and cathode sides, which consist of a sintered stainless steel sheet and a hydrophobic carbon cloth. The sintered stainless steel sheet provides mechanical support to sustain the imposed pressure difference between the anode and cathode, while hydrophobic carbon cloth has multiple purposes. It prevents water flooding, especially on the anode side, reduces the contact resistance between the catalyst-coated membrane and the sintered stainless steel sheet, and minimizes the damage to the catalyst-coated membrane by the hard fibers of the sintered stainless steel sheet

[32]. In the case of water buildup, while continuously providing high relative humidity of PEM, there is a risk of anode flooding, which decreases EHC performance due to the reduced active area. Alongside water input via saturated hydrogen, water also transfers to the anode via a diffusion process from the cathode back to the anode due to the pressure difference between the two-compartment areas. For this reason, carbon-based GDL can be treated with hydrophobic polytetrafluoroethylene (PTFE) dispersion. However, anode flooding presents a challenge for EHCs that contain metallic GDL, as PTFE treatment cannot be applied due to PTFE particles blocking the pores (porosity $<40\%$) and their inadequate adherence to the surface of the pores [33]. Regarding EHC application, the chosen material for GDLs should have high electrical conductivity, good mechanical strength under high pressures, suitable porosity, good chemical stability, and corrosion resistance, as well as the ability to effectively enable water transfer.

2.2.4 MEA Gasket

As mentioned before, PEM in combination with CLs and GDLs forms a layered MEA. The MEA directly affects the performance of EHC and it's the most expensive and fragile part. For this reason, it is of utmost importance to ensure proper gasketing for it. There are multiple ways to effectively form a sealing structure of the MEA. For example, the five-layered MEA gasket includes an elastic seal as a contact between flow field plates and resin frame connected with adhesive to PEM. The primary regions of an assembly important for gasket incorporation can be distinguished. From the gasket point of view, the MEA can be classified into three regions: function region (active area), structure region, and transition region. The function region contains the active area of the EHC, consisting of electrodes and membranes. The transition region is located between the active area and the structure region and contains the elements of both as well as the gap in between. This region has to be constructed to ensure there is no hydrogen leakage. For elongated lifetime and better performance, the gap can be overlapped, breached, or glued, but it must not be the cause of mechanical deformations to the MEA [34]. It should also be carefully designed as an effective seal due to the highly pressurized cathode compartment, which presents a risk of hydrogen leakage. The structure region has to provide mechanical stability and durability for the MEA to protect the PEM and additionally ensure sealing. Also, it acts as an electrical insulator and can, with the thickness of the gasket, play the role of a distance keeper that controls the amount of contact between flow field plates and MEA. This contact (a dimensional mismatch between the gasket and GDL) is the deciding factor for the amount of contact resistance, which greatly influences the EHC performance. For this reason, it is important to precisely match the thicknesses of the gasket and the GDL to have optimum conditions for EHC operation. The potential mismatches due to thinner or thicker gaskets may lead to various problems. A thinner gasket causes sealing problems and hydrogen leakage, as well as higher mass transport losses, while a thicker gasket results in a gap between the flow field plate/bipolar plate and the GDL, causing higher contact resistance, which negatively reflects on the ohmic overvoltage [35]. For these reasons, the optimum contact between the gasket and the GDL of EHC should be achieved.

2.2.5 Flow Field Plates

Flow field plates, especially in large cell-numbered stacks, physically make up a big part of the assembly. Therefore, the materials from which they are made must be good electrical conductors, corrosion resistant, and low weight. In PEMFCs, flow field plates constitute more than 60% of the weight and 30% of the total cost in a stack. Also, an optimal design of the bipolar plate can improve the overall stack performance even to 50%. Some of the possible channel configurations for flow field design are pin-type, parallel, crisscross, single serpentine, multiple serpentine (parallel serpentine), mixed serpentine, serially-linked (subsequent) serpentine, mirror serpentine, interdigitated, biomimetic (fractal) interdigitated, spiral interdigitated, metal mesh (screen), metal foam (porous media), porous plate, tapered, constricted, and stepped (asymmetric channels), radial, structural-based, and integrated. The dimensions and shapes of these designs influence hydrogen flow direction, pressure drop, water removal, mechanical stability, heat and mass transfer, as well as electrical conductivity and resistivity [36]. For all these facts, a parallel with EHC stacks can be drawn, as previously mentioned, their design is similar to PEMFCs. The most used materials for flow field plates are graphite, alloys, or metals such as stainless steel or Ti. The flow field's primary role in the assembly is the distribution of reactant (hydrogen) to the electrochemically active areas, where they also uniformly distribute gas and current, remove excess heat, and mechanically support EHC from high-pressure differences. Furthermore, the shape of the flow field has to provide strong support to the MEA, but with a low-pressure drop. Researchers and engineers continue to explore new materials and manufacturing techniques to improve the performance and cost-effectiveness of EHC flow field plates.

2.2.6 End Plates

End plates play a crucial role in providing structural support, sealing, and facilitating the flow of reactants and products within the cell and a whole stack. They are clamping parts on the outside, forming the EHC assembly, providing mechanical strength and a compressive force that ensures good electrical contact between inside layers. The optimal clamping pressure should be tested to ensure the best performance. The generally selected material for endplates is stainless steel (a metal material), as it can resist high cathode pressures and temperatures. Nonetheless, high enough pressures could deform the endplates at the centerline of the stack, if they are not properly sealed and the stress is not uniformly distributed [2]. The end plates also help create a seal to prevent the leakage of reactant, which is crucial to maintaining the efficiency and safety of the EHC system. The sealing formed at various interfaces prevents hydrogen leakage. Furthermore, the end plates serve as a rigid structure to support the individual cells within the EHC stack. It helps maintain the mechanical integrity of the stack and ensures that the individual cells are properly aligned and compressed to maximize performance. In stacks, the plates typically contain channels and/or flow fields that distribute hydrogen gases to the respective electrodes (anode and cathode) of each cell. In some designs, the end plate may also have channels for the circulation of coolant. This helps regulate the temperature within the stack, as maintaining an optimal operating temperature is essential for efficiency and durability.

2.2.7 Stack Configurations

Industrial EHC systems consist of multiple EHC cells in a stack configuration to achieve higher hydrogen flowrates and gas pressures as single-cell performance can be affected by hydrogen molecule back-diffusion of the gas across the membrane, which is exacerbated at high pressures. There are different stack configurations based on the series or parallel connection of hydrogen flow channels, but all cells are electrically connected into a series so that the total applied voltage to the stack is equally divided by each cell. The flow in parallel configuration is compatible with EHC applications where higher gas throughput is needed at limited pressures, whereas the flow in series configuration can deliver higher pressures but at lower hydrogen flowrates. Furthermore, if optimization between delivered flow and pressure is required, it is also possible to design a hybrid configuration that combines both parallel and series flow aspects depending on a needed application. For the same number of cells and operating conditions, the efficiency for the stack with parallel flow is higher than the series flow or hybrid stack. However, the parallel flow configuration is more susceptible to stack failure as all PEMs are subjected to a high-pressure differential so single membrane failure will instantaneously depressurize and disable the entire stack. On the other hand, membrane failure is less likely in the flow in series configuration since the overall pressure difference is divided across all the PEMs in series [37].

2.3 General Operating Principle

EHCs can be used for hydrogen compression, hydrogen purification, or a combination of the two processes, as in the middle of an EHC is a hydrogen-selective proton-exchange membrane usually composed of Nafion—a conductive polymeric membrane. For this reason, the inert gas components cannot cross the membrane, and the EHC performs as a purifier of hydrogen gas.

The start of the EHC work is characterized by applying the relatively low potential difference (EHC is more suitable for low-power applications—more will be said in Sect. 5.1) at the external circuit and feeding the anode inlet with humidified hydrogen or other gas-hydrogen mixture. The operating temperature of the inlet stream should be high enough to suppress any mass transfer resistance of hydrogen gas through the GDLs and the catalyst layers, as well as to maintain the membrane fully hydrated. Hydrogen feed at the anode is oxidized to protons (hydrogen molecule splits into two protons, H^+ and two electrons, e^- as presented in Eq. (1)) while remaining gases cannot pass through the selective membrane and are therefore collected at the anode outlet.

Hydrogen reaction at the anode (1):

$$[HOR] : H_2(low - pressure) \rightarrow 2H^+ + 2e^- \tag{1}$$

The hydrogen oxidation reaction (HOR) occurs due to the applied potential difference and catalyst influence at the three-phase boundary (TPB), which consists of a proton conductor (electrolyte Nafion) for hydrogen protons, an electron conductor (carbon/metal material of GDL) for electron travel, and a virtual "porosity" phase for transporting hydrogen and water molecules. A catalyst in the EHC is required to increase the rate of the catalytic reaction. The most often used catalyst material to catalyze the HOR and hydrogen reduction reaction (HER) for both anode and cathode, respectively, is platinum (Pt). However, the Pt is not economically feasible due to low surface areas, which is why a support material is required to obtain a high dispersion and a narrow distribution of catalysts that can also interact and influence the catalytic activity. Therefore, Pt is often used as a catalyst in the form of Pt/C, i.e., Pt nanoparticles on an activated black carbon support (C). This form presents an optimized use of Pt to enable a high surface area-to-volume ratio and activity [38]. After HOR on the anode, electrons pass through the electrically conductive black carbon particles, GDL fibers, and bipolar plate or other electrical collector to an external circuit towards the equal corresponding components of the cathode, converging with the hydrogen protons transported through the Nafion membrane for HER (two hydrogen protons, H^+ and two electrons, e^- form a hydrogen molecule as presented in Eq. (2)) to form the hydrogen molecule again at the cathode.

Hydrogen reaction at the cathode (2):

$$[HER] : 2H^+ + 2e^- \rightarrow H_2(high - pressure) \tag{2}$$

After HER, the reduced hydrogen molecules take place in the confined cathode volume compartment of the flow field plate at elevated pressure, where they can be discharged at high pressure through the hydrogen outlet. The elevated pressure is achieved via an electric current that is continuously pumping more hydrogen (as seen in (2), two electrons with two protons form one hydrogen molecule) from the anode to the cathode side, where outlet flow is restricted by a back-pressure regulator.

Total hydrogen reaction in the EHC (3):

$$H_2(low - pressure) \rightarrow H_2(high - pressure) \tag{3}$$

Therefore, the pressure can be relieved with the regulator, and on the other hand, in theory, it can increase as long as the electric current is applied. However, if there is no leakage, the achieved hydrogen pressure is limited by the EHC design and materials, i.e., mainly membrane deformation under high-pressure differences, its proton conductivity and permeability for hydrogen molecule back diffusion, and mechanical strength. The time needed to compress hydrogen is influenced by the supplied current density and the volume of hydrogen's confined compartment.

2.3.1 Hydrogen Purification via EHC

As mentioned before, purification via EHCs is based on the high selectivity and low permeability characteristics of PEM. When an anode inlet stream contains a gas-hydrogen mixture, high-purity hydrogen (up to 99.999 vol.%) is formed at the cathode outlet, with current controlling the hydrogen separation rate (Faraday's Law). High hydrogen separation with high efficiency can be achieved at low cell voltages [26]. The anode side in EHC has to oxidize hydrogen in the presence of impurities that have a negative effect on the HOR performance (the high HOR current density has to be maintained). Thus, if EHC is utilized as a purifier, the use of alternative catalysts to pure Pt nanoparticles is required for efficient hydrogen separation. As one of the main objectives of the EHC as a purifier is to separate pure hydrogen from an impure gas stream made via steam reforming, oxygen-blown autothermal reformer (ATR), or coal gasification, the feed stream that EHC has to purify will contain the presence of CO, CO_2, methane (CH_4), inert gases, and other impurities such as ammonia (NH_3) and hydrogen sulfide (H_2S) [39].

Reformate hydrogen can contain up to 3% CH_4, while if that inlet comes directly from natural gas (in which providers usually insert portions of hydrogen), 90–95% CH_4 can be expected. It was measured that 80% H_2 recovery is possible with electrochemical purification in a 25% H_2/CH_4 gas mixture [40]. Regarding Pt-based electrocatalysts, CH_4 seems to act mainly as a diluent for hydrogen, similar to inert gases. An issue regarding EHC purification operations that can occur in a dead-ended anode is fuel starvation, when inert gas (e.g., N_2) with a combination of liquid H_2O, locally depletes the hydrogen content in the anode and has a negative effect on the cell's performance and durability. This phenomenon particularly presents a problem when high currents are applied and hydrogen content in the input stream is below 50%. Hence, hydrogen inlet should not contain more than 6% inert gases to enable the targeted performances ($j > 1$ A/cm^2). In addition, if a dead-end or multi-stage purification/compression method is chosen, larger dilution effects have to be considered. In the case of H_2-NH_3 gas mixtures, when the anode is fueled by pure hydrogen after a sequence of NH_3 poisoning, the recovery does not depend on the amount of NH_3 previously supplied, but on the time of that supply [39]. A short-term poisoning, even with a concentration as high as 200 ppm, can be recovered. On the other hand, a long-term poisoning with only 5 ppm of ammonia, cannot be fully recovered as, after some time, the contact between the Pt nanoparticles and the ionomer is lost, leading to an irreversible loss of active surface area [41].

CO_2 presents one of the main impurities when dealing with hydrogen reformate, and its behavior with hydrogen in the presence of Pt electrocatalyst can be described as a reverse water gas shift (RWGS) reaction [42]. The RWGS reaction is a chemical reaction that involves the conversion of CO_2 and H_2 to produce CO and H_2O. The direction of the reaction can be influenced by factors such as temperature, pressure, and the concentrations of reactants and products. The equilibrium of the RWGS reaction can be shifted by adjusting the reaction conditions. For example, increasing the temperature generally favors the formation of carbon monoxide and water, while lower temperatures favor the formation of carbon dioxide and hydrogen. The choice of catalysts can also influence the reaction rates and selectivity [43]. The reaction is represented by the following Eq. (4):

$$CO_2 + H_2 \rightleftharpoons CO + H_2O \tag{4}$$

That reaction between CO_2 and H_2 is believed to occur in two places. First, directly in the gas phase, when both gases are in contact, and second in the presence of the Pt catalyst, where CO_2 reacts with hydrogen adsorbed protons (H_{ad}) to produce CO molecules that easily poison the Pt surface [44]. The phenomena can be described with two following reactions (5) and (6):

$$2Pt + H_2 \rightarrow 2Pt - H \tag{5}$$

$$CO_2 + 2Pt - H \rightarrow Pt - CO + H_2O + Pt \tag{6}$$

CO is a very common impurity present in hydrogen reformate that has a critical impact on Pt catalyst activity in a way that it adsorbs very strongly on Pt, as seen in Eqs. (7), (8), and (9), thus, blocking the catalyst sites for other reactions such as H_2 dissociation and oxidation. H_2S can be present in proportions up to 10 ppm in natural gas and, similar to CO, can strongly adsorb on Pt surfaces, which blocks catalyst active sites and reduces the anode performances and therefore the cell voltage [39].

$$CO + Pt \rightarrow Pt - CO \tag{7}$$

$$Pt + H_2O \rightarrow Pt - OH + H^+ + e^- \tag{8}$$

$$2Pt - OH + Pt - CO \rightarrow 3Pt + CO_2 + H_2O \tag{9}$$

An experiment examined the feasibility of a single-cell EHC when separating hydrogen from a reformate stream that is derived from a hydrocarbon source such as natural gas or an alcohol such as methanol. The separation of both pure hydrogen and impure hydrogen containing 1000 ppm CO and CO_2 was analyzed. With pure hydrogen, the EHC efficiency is quite high (greater than 80% with about 80% hydrogen extraction) for the extraction of hydrogen from a hydrogen/CO_2 mixture. However, when the hydrogen contains 1000 ppm CO, the efficiency is relatively poor as hydrogen-containing CO severely poisons the anode and raises the anode potential (by as much as 300 mV) [45]. Hydrogen recovery can be increased through the use of one or more additional EHC cells in series with some energy penalty, or more poison-tolerant anode catalysts. When investigating EHC for the purification of binary H_2–CO_2 mixtures, it was noted that an electrochemical reduction of CO_2, responsible for CO formation, may occur on the catalyst, hindering the hydrogen splitting and subsequent permeation through the membrane [31]. Therefore, a catalyst on the anode side that operates with impurities alongside hydrogen, must demonstrate very high HOR activity and be tolerant to the poison molecules, as mentioned before. In that regard, Pt is usually alloyed with a less noble catalyst. By changing the Pt electronic structure, the Pt–CO bond is weakened, and components adsorbed on less noble metal sites can diffuse toward Pt sites. For example, PtRu/C anode catalyst can be considered experiments showed it is more poison tolerant than Pt/C and could therefore be used for the subsequently poisoned feed [46, 47]. On the other hand, Pt/C was shown to be a very effective catalyst for the heterogeneous hydrogenation of acetylene [48]. Also, due to temperature increasing the rate of CO oxidation, operating EHC at higher temperatures can reduce the poisoning effect, but likely at the expense of the electrocatalyst's durability [49]. Pressure seems to have only a slight amelioration effect on the poisoning, while for a higher current density, the cell potential decreases much faster [50]. EHC demonstrated high hydrogen purities despite CO concentrations up to 12%, and even H_2S [51].

2.3.2 Water Management: Input Hydrogen Humidification

For hydrogen protons to travel across the PEM to the cathode side of the EHC, the membrane needs to be humidified for efficient proton conduction and to prevent membrane dehydration due to temperature gradients forming during EHC operation. The electrochemical compression process produces heat mostly throughout the MEA, owing to the resistance to proton transport within the PEM, the reversible heat of reaction, and the irreversible heat production caused by the electrochemical reactions [52]. Input hydrogen with a low relative humidity causes unstable EHC operating conditions. The local dehydration of the PEM due to the unbalanced water transport across the membrane increases its resistance. For this reason, a higher voltage has to be supplied to the EHC to allow for good performance, which decreases its overall efficiency [7]. Therefore, EHC in hydrogen systems is often incorporated after a PEM electrolyzer (PEMEL), which produces already water-saturated hydrogen. Otherwise, a humidifier should be installed in the system before hydrogen EHC input. In order to maintain high ionic conductivity in the membrane, the input hydrogen (or hydrogen-gas mixtures) is usually humidified by passing through a bubbler type of humidifier before entering the cell. It generally consists of a heated bottle filled with distilled water through which the gases are bubbled. The bottle can be heated by a magnetic hotplate stirrer. This kind of humidifier is generally held at a somewhat higher temperature than the cell temperature to get full humidification of the feed gas. The intake gas can be determined by a flow meter and bubbler to control the flow rate and inlet temperature, respectively. The temperature is set on the heating plate of the bubbler to enable the gases in the upper part of the bottle to have the same temperature as the hydrogen gas input line (the line can be heat traced to maintain the required temperature). It allows an easier water uptake by the gases, avoiding an excessive cooling of the flow passing through the water [53]. The partial pressure of water vapor in hydrogen (and therefore the relative humidity) can be adjusted by setting the temperature of the humidifier. The hydrogen flow across the gas humidifier should be sufficiently low to reach equilibrium. Therefore, the water vapor concentration in hydrogen is assumed to be close to the saturation partial pressure at the temperature of operation of the humidifier [54]. Another option for EHC humidifying instead of external water management is an internal water compartment incorporated into the EHC design. For example, water management can rely on a water chamber built into the cathode plate. Due to the induced pressure gradient between the cathode and anode during EHC operation, water flows from the cathode to the anode, hydrating the PEM. The water flow driven by the pressure gradient can compensate for the electro-osmosis flow from the anode to the cathode [55].

2.3.3 Thermal Management: Cooling System

Although the hydrogen compression via EHC is isothermal, cell resistances generate losses in the form of displaced excess heat, causing temperature increase and uneven distribution throughout all components. Accumulated heat transfers from one EHC component layer to the other by conduction, convection, and radiation. The main heat source is located in PEM due to (ohmic) resistance to proton transport within, as a consequence of the Joule effect. Even in single-cell EHC, increasing the applied potential difference is followed by the ohmic resistance and Joule's heat due to the electrical current increase, which causes the PEM's temperature to rise [56]. If the EHC is formed as a stack, every cell generates heat to the adjacent cells on the right and left sides that are connected. This causes the heat to build up, leading to a significant increase in operating temperature, causing the decrease of water content in PEM, which can render it inoperative. Therefore, the cooling system must be applied to EHC due to the accumulated heat, especially at high operating current densities. The overvoltages in the cells generate heat, which must be removed to maintain constant EHC performance and to avoid temperature-induced variations between cell components, as well as between cells in a stack [57]. An example of cooling for EHC is convection cooling via the perimeter of the stack. Assuming that heat is generated uniformly over the active area, a convection coefficient can be estimated for free convection. Also, the effective convective surface area can be increased with just the ends of the cells being exposed for cooling via the cooling fins that protrude between cells [58]. In the case of an increased need for cooling, the

EHC can be thermally controlled via a water-circulating system mounted on the anode and cathode plates, to generate a homogeneous temperature profile along the channel. The MEA and GDL can be sandwiched between two separators grooved by channels, while constant temperature water is supplied to the separator to keep the cell temperature constant. Furthermore, the hydrogen gas lines can be thermostated at higher temperatures to prevent water condensation.

2.3.4 Water Management: Output Hydrogen Dehumidification

As PEM needs to be hydrated for efficient proton conduction, the resulting hydrogen gas is saturated with water. The water molecules transfer through PEM (both from the anode to the cathode and vice versa) as a result of electro-osmosis, hydraulic permeation, thermo-osmosis, and diffusion mechanisms. The actuators of those water transport phenomena are local gradients of electrochemical potential, pressure, temperature, and concentration, which cause water back diffusion from the cathode to the anode [59]. On the other hand, water transfer from the anode to the cathode side is due to electroosmosis, i.e., water molecules being dragged across the PEM by the hydrogen protons due to their mutual electrostatic draw [60]. Having humidified output hydrogen from the EHC can be hindering, depending on the afterward hydrogen application. In Europe, if considering hydrogen at the dispenser nozzle used for road vehicles with PEMFC, part of the ISO standard 14687-2 states that the maximum permissible concentration of water in hydrogen should not exceed 5 $\mu mol_{H2O}/mol_{H2}$ [61]. Although EHC's output hydrogen is humidified, one of the effects observed during isothermal compression is hydrogen partial dehumidification. It is similar to 'air conditioning' effects when the pressure is increased and subsequently released. In this way, compression of hydrogen gas directly increases its pressure, but compression of water vapor leads to condensation of surplus water as the water vapor pressure is determined by local dewpoint temperature. In case of a positive pressure differential, and isothermal conditions, not all water carried into the anode side can evaporate on the (over-) saturated cathode side [62]. Hence, less water vapor exits the EHC after compression, but not sufficiently enough to comply with the mentioned standard or to prevent failure of water-sensitive equipment such as various sensors and indicators (e.g., flow meter, hydrogen purity indicator, pressure sensor), avoiding water condensation in them. Therefore, as a part of the EHC system, before the hydrogen is stored in a high-pressure vessel, a dehumidifier should be installed that can withstand appropriately high pressures. For example, hydrogen at the cathode outlet can be sent to a condenser, where it is cooled using ice in order to condense the water present in the gas steam. The water that is condensed out of the hydrogen cathode outlet stream can be held in a storage tank, and then filtered and treated for reuse upstream to humidify the anode inlet stream.

To sum up, the most important requirements for efficient EHC system operation are:

- water management (humidification and dehumidification of hydrogen),
- thermal management, which depends the most on the number of EHC cells (required cooling and limiting operating current density),
- design modifications to effectively withstand high pressures (mechanical strength, efficient sealing for possible hydrogen leakage, prevention of membrane deformation, hydrogen back diffusion, and contact resistances),
- necessary anode poison-tolerance, relevant to the hydrogen-gas mixture input stream for EHC purification, and
- the insurance of safe EHC operation/handling (prevention of damages due to high pressures and hydrogen leakage, staff education).

3 Electrochemical Hydrogen Compressor Analysis

In this chapter, EHC will be analyzed regarding its performance and possible applications. EHC operation is significantly influenced by various overvoltages, which affect its efficiency. Also, the optimal operating conditions and the total cost of the system will be discussed. To conclude, the importance of the safety aspect of EHC operation will be highlighted.

3.1 Performance

As discussed in Sect. 3, currently, hydrogen compression is mostly achieved via mechanical compressors. This compression process is adiabatic and hence divided into stages to reduce temperature increases and improve energy efficiency. Typically, the friction of seals and valves adds up to 20% to the energy requirement, resulting in approx. 6 kWh/kg of energy is needed for compression from 10 to 400 bar. The EHC compression process is isothermal and single-stage, having the potential to reduce this energy requirement for the same pressure increase down to 3 kWh/kg [51]. The higher the hydrogen pressure is achieved; the higher compression energy is needed. Therefore, it is worth mentioning that for a final pressure of 800 bar, the compression energy requirements for isothermal compression ($\eta = 50\%$) would amount to about 11.5% of the energy content of hydrogen (higher heating value (HHV)), while for the ideal adiabatic compression, it would amount to 15%. The basic adiabatic thermodynamic process ($s = \text{const.}$, $\varphi_{1-2} = 0$) of specific work done by a mechanical compressor is described using the Eq. (10):

$$w_{adiabatic} = \frac{\kappa}{\kappa - 1} p_1 v_1 [1 - (\frac{p_2}{p_1})^{\frac{\kappa-1}{\kappa}}] = \frac{\kappa}{\kappa - 1} R T_1 [1 - (\frac{p_2}{p_1})^{\frac{\kappa-1}{\kappa}}] \tag{10}$$

$w_{\text{adiabatic}}$ (J/kg) is the specific compression work, p_1 (Pa) and p_2 (Pa) are, respectively, the initial and final pressures, v_1 (m^3/kg) is the initial specific volume, T_1

(K) is the initial temperature, $R = 8.314$ J/mol K is the ideal gas constant, s (J/kg) is specific entropy, and κ is the ratio of specific heats, also known as the adiabatic coefficient, that for hydrogen gas amounts to 1.41. Besides Eq. (10), as already mentioned, the actual compression work of a mechanical compressor is usually larger than the one calculated, owing to additional losses, such as flow and friction losses. Compared to adiabatic compressor performance evaluation, the basic isothermal thermodynamic process (T = const., pv = const.) of EHC compression-specific work, $w_{\text{isothermal}}$ (J/kg) follows a simpler Eq. (11).

$$w_{isothermal} = pvln\frac{p_1}{p_2} = RTln\frac{p_1}{p_2} \tag{11}$$

The same result is derived from the Nernst equation for the pressure electrolysis of water, as seen in Eq. (12). In both cases, the compression work is the difference between the final and the initial energy states of the hydrogen gas [63]. Nernst voltage, E_{N} (V) is the theoretical minimum voltage that needs to be applied to achieve the wanted hydrogen pressure. It depends on the standard cell potential, E_0 (V) which is equal to zero for EHC as the cathode reaction is the reverse of the anode reaction, number of hydrogen electrons, $z = 2$, Faraday's constant, $F = 96{,}485$ C/mol, final (cathode) hydrogen pressure, p_{c} (Pa) and initial (anode) hydrogen pressure, p_{a} (Pa) ratio, and EHC operating temperature, T (K) [64].

$$E_{Nernst} = E_0 + \frac{RT}{zF}ln\frac{p_c}{p_a} \tag{12}$$

By considering the deviations of the effective cell voltage in the form of overvoltages, an overall cell voltage can be written as Eq. (13). Due to the additional resistances that occur during EHC operation, its actual compression work is usually larger than the one calculated in Eq. (11) [30].

$$E_{EHC} = E_{Nernst} + \eta_{act} + \eta_{Ohm} + \eta_{mass-transf} + \eta_{diff} \tag{13}$$

These kinetics limitations increase when the current density increases, which means that the efficiency of an EHC will decrease when its productivity (flow rate of pure H_2) increases. The overall (applied) EHC voltage depends on the:

- The ohmic resistance of the cell, η_{Ohm} (V) both owing to the non-infinite conductivity of the PEM and electrode materials and the uneven electrical contacts between EHC layers,
- charge-transfer limitations in the form of polarization (activation) overvoltage, η_{act} (V) even though the HER and HOR are fast reactions on Pt-based electrodes, however, their overvoltage values are strictly not zero,
- mass-transport limitations, $\eta_{\text{mass ? transf}}$ (V) due to molecular hydrogen and proton transport to/from the catalytic sites,

- and diffusion overvoltages, η_{diff} (V) due to hydrogen gas crossover through the membrane [65].

Mass-transfer overvoltage includes flow limitations that hydrogen encounters when approaching the catalyst due to high current inserting at a faster pace. Therefore, at higher current densities, mass-transfer overpotential is getting more pronounced as the membrane's water content is lowered [41]. Ohmic overpotential includes resistances associated with electronic resistance of other electrically conductive components (e.g., current collectors, bipolar plates, GDLs.), interfaces between the various layers, as well as resistance to proton transfer through PEM (Eq. (22)) [66]. PEM resistance is determined by the thickness, φ (cm) and the relative humidity of the membrane, σ_m (S/cm) with linear relation to current density, i (A/cm^2) as shown in Eq. (23) [67]. It represents the biggest part of the ohmic voltage loss, which is why it is usually the only one calculated.

$$E_{Ohmic} = R \times i \tag{14}$$

$$E_{Ohmic} = \frac{\varphi}{\sigma_m} i \tag{15}$$

The membrane conductivity is influenced by the operating temperature, T (K), and water content (degree of humidification), λ of the membrane. It can be calculated in multiple ways, e.g., using a simple equation where σ_m (S/cm) is a linear function of temperature, T (°C), (12) [68], or semi-empirical expression with the influence of water content (13) [69], which is usually taken from the literature experiments for specific PEM type or can be calculated by experimental correlations, e.g., depending on the water vapor activity [67]. Generally, excluding PEM in a dry state, the value of λ for Nafion membrane is in the range of 2–14, depending on the level of the membrane exposure to solely water-saturated gas, or 14–22 if it is also humidified with liquid water, with 22 being only in contact with liquid [70].

$$\sigma_m = 0.0007T + 0.0574 \tag{16}$$

$$\sigma_m = (0.005139\lambda - 0.00326)^{1268\frac{1}{303} - \frac{1}{T}} \tag{17}$$

The polarization overpotential is the sum of anode and cathode overpotentials (Eq. (18)), which describe the kinetics of the occurring electrochemical reactions. They describe the resistance that must be overcome by an electrochemical reaction before the reaction can occur and is calculated by the sum of Tafel equations (19) and (20), where T (K) is the operating temperature, i (A/cm^2) is the operating current density, α is a charge transfer coefficient, assumed equal to the symmetry factor (i.e., 0.5) for both electrodes, and i_0 (A/cm^2) is an exchange current density that presents a measure of the electrode's electrochemical activity before the application of potentials [67].

$$\eta_{act} = \eta_a + \eta_c \tag{18}$$

$$\eta_a = \frac{RT}{zF\alpha_a} ln(\frac{i}{i_{0,a}}) \tag{19}$$

$$\eta_c = \frac{RT}{zF\alpha_c} ln(\frac{-i}{i_{0,c}}) \tag{20}$$

The overall EHC efficiency, η_{EHC} (Eq. (21)) should include voltage efficiency, η_{voltage} (Eq. (22)) that shows the effects of cell voltage losses through theoretical (Nernst voltage) and applied cell voltage, as well as current (Faraday) efficiency, η_{current} (Eq. (23)) that expresses the faradaic losses due to the hydrogen molecule back-diffusion. It can be emphasized that the higher the EHC operating temperature, the lower the current efficiency since the electrolyte resistance is lower, leading up to higher crossover currents [71].

$$\eta_{\text{EHC}} = \eta_{\text{voltage}} \times \eta_{\text{current}} \tag{21}$$

$$\eta_{\text{voltage}} = \frac{E_{\text{Nernst}}}{E_{\text{EHC}}} \tag{22}$$

$$\eta_{\text{current}} = \frac{n_{\text{output}}}{n_{\text{forward}}} \tag{23}$$

Current efficiency describes EHC's mechanical reliability and is affected by the hydrogen back-diffusion and possible leakage on the cathode side. Therefore, it can also be termed the "flux efficiency" and be described through the ratio of output hydrogen mols, n_{output} (mol) and forward hydrogen mols, n_{forward} (mol) as seen in Eq. (23). Molar hydrogen flow from the anode input, towards the cathode output, follows Faraday's law (Eq. (25)), and is determined by the applied current, I (A), number of electrons, and Faraday's constant [30], while hydrogen backward molar flow rate, $q_{n,\text{ backward}}$ (mol/s) depends on the permeability constant, DH (mol/(Pa × m × s)), active area, A (cm^2), membrane thickness, φ (cm), and the pressure difference causing the hydrogen flow from cathode to the anode as seen in Eq. (26). The permeability constant, DH depends on the nature of the hydrogen gas and membrane characteristics [30]. Output hydrogen molar flux is the sum of forward and backward hydrogen flow (Eq. (24)).

$$\eta_{EHC} = \eta_{voltage} \times \eta_{current} \tag{24}$$

$$\eta_{voltage} = \frac{E_{Nernst}}{E_{EHC}} \tag{25}$$

$$\eta_{current} = \frac{n_{output}}{n_{forward}} \tag{26}$$

When voltage and current efficiency are considered, the EHC efficiency of around 60% can be obtained at low pressure (<10 MPa) and with an operating current density of around 1 A/cm^2. Even if the EHC total efficiency is lower than the electrochemical compression efficiency (isothermal efficiency), the EHC can be considered highly efficient in comparison with mechanical compressors [7].

3.1.1 Operating Conditions

Electrochemical compression is more efficient at low-pressure operation as high-pressure conditions are not beneficial due to the increase of the hydrogen molecule permeation across the membrane from the high-pressure cathodic compartment to the low-pressure anodic compartment (hydrogen back-diffusion), with a resulting decrease in the Faradaic contribution. Furthermore, molecular hydrogen back-diffusion also impacts hydrogen purity. Several strategies of mitigation can be considered: adding a molecular hydrogen barrier or a modifying PEM, avoiding excessive pressure differences on either side of a membrane and using instead a stack (several steps of (intermediate) compression), and optimizing current density and water management to avoid permeate backflow and electrical resistance [57]. Another important issue regarding single-cell EHC high-pressure operation is the mechanical deformation of PEM that can occur due to a lack of membrane physical support or inadequate support.

Also, galvanostatic and tensiostatic modes of operation have been investigated. In the first case, one measures the cell voltage (driving force) for an assigned current density (rate of the process), which is related to the rate of hydrogen production in the cathodic compartment, while in the second case, one measures the current density for a given value of applied voltage. In other words, with galvanostatic conditions, the rate of the process is assigned, and the driving force is measured, while under tensiostatic conditions, the driving force is assigned and the rate of hydrogen production is measured. In the galvanostatic case, erratic behavior of the cell has been observed, mainly because of the non-controllable variations of the membrane water content. Therefore, tensiostatic operating conditions are preferred to galvanostatic ones, since the former allows self-control of the process kinetics, according to the hydration condition of the membrane. On the contrary, by fixing the kinetics of the process, as it occurs under galvanostatic conditions, the water degree of the membrane exhibits an erratic behavior [72].

The operating conditions of the system in the form of input parameters such as active area, A, operating temperature, T, pressure gradient Δp, current density, i, and total applied cell voltage U_{EHC}, depend on the purpose of the EHC utilization. For compression application, the optimal energy consumption is obtained for a cell current density of $i = 0.5$ A/cm^2, and the corresponding cell voltage close to $U_{EHC} = 0.2$ V, with cell resistance of about 0.3 Ωcm^2 $< R_{cell} <$ 0.4 Ωcm^2, and hydrogen pressure gradient, Δp between anode and cathode volumes of 50 bars at EHC operating temperature, T at 80 °C. For a specific hydrogen flow to be achieved, the active area, A is chosen regarding the operating current density (cell voltage, which directly depends on the applied current density) and the MEA resistivity. On the other hand, for additional purification, the required current density shall reach approx. $i = 2$ A/cm^2 to compete with the existing (non-electrochemical) membrane compression systems, and the cell voltage shall be maintained at a reasonably low value of below 1 V [65].

3.2 Applications

Efficient hydrogen separation and compression technology must be developed to realize the future hydrogen economy. The EHC converts low-pressure and impure hydrogen gas to high-pressure hydrogen gas with high purity. Considering EHC advantages and disadvantages, multiple potential EHC applications are going to be discussed in more detail in this chapter, such as hydrogen separation from the natural gas grid, hydrogen purification after hydrogen production from fossil fuels, hydrogen compression for high-pressure storage vessels and transportation, hydrogen purification for its use in PEMFC and other applications that require high hydrogen purity and/or high hydrogen pressures (even up to 1000 bar [57]).

3.2.1 EHC Advantages and Disadvantages

With the absence of mechanical losses, EHC belongs to the category of a solid-state compressor, ensuring compatibility and a longer life, which is expected to reduce the cost of operation and maintenance. Moreover, the EHC stack with multiple cells seems to be more advantageous than a multistage mechanical one as it can reach the same pressure level as a mechanical one using fewer stages, with no noise produced and a simpler design [73]. Also, the required hydrogen output flow rate and output pressure can be regulated by arranging cells in series or parallels. Hence, EHC can be more reliable and preferable for pressurizing hydrogen gas over mechanical compressors, especially due to higher efficiency, as explained in Sect. 5.1 [74]. In terms of purification application, the use of EHC could be beneficial in numerous industrial fields, such as hydrogen purification for the use in fuel cell technologies, cooling agents in turbines, for the application in nuclear reactors (separation and concentration of hydrogen isotopes), and separation of hydrogen from natural gas mixtures [26]. EHC's main advantages are:

- low operation and maintenance cost (OPEX),
- high-purity hydrogen production (up to 99.999%),
- high tolerance to impurities,
- high hydrogen recovery (up to 99%),
- no moving parts,
- no vibration and noise,
- modularity (connecting cells in an EHC stack),
- simultaneous one-stage hydrogen separation and compression,
- high compression efficiency.

while its main disadvantages include:

- difficulty in manufacturing the cell assembly,
- difficulty in realizing a perfect sealing,
- the need for additional water and thermal management (which also increases the capital expenditure (CAPEX) of a whole EHC system),
- high cell resistance,
- hydrogen back diffusion,
- membrane deformation due to high-pressure differences [7].

3.2.2 Total Cost of EHC System

Cost assumptions of the CAPEX for EHC are approx. \$3850/kg/day, compared to the \$6050/kg/day of a mechanical compressor, while the total cost of ownership (TCO) for the same amounts to \$3.44/kg, and \$4.77/kg, respectively [75]. The OPEX is also found to be lower (<\$1/kg for EHC, and \$1.75–2.3/kg for mechanical compressor) due to the lower specific energy consumption, generally <4 kWh/kg [7]. Furthermore, predictions are that OPEX of \$3400/year and total cost of \$0.503/kg can be achieved due to several improvements achieved in recent years, such as optimization of the catalyst, enhanced thermal management, improved water management, and the use of low-cost membranes. The capital costs of EHC include 79.8% of a stack and 20.2% of the balance of the plant (BOP). The main expenses of the EHC stack consist of the following costs: MEAs (40.2% of total CAPEX), cell frames (14.1%), end plates (9.5%), and MEA support (7.1%) [76]. A proper design of EHC allows for an operating lifetime of 10 years, and even 20 years [76], and a mean time between failures (MTBF) as high as >40,000 h [75].

3.2.3 Hydrogen Blending into Natural Gas Grid

Hydrogen is an important energy vector that can be produced from RES during off-peak hours, stored, and distributed by injecting into existing natural grids. Blending hydrogen into the natural gas pipeline is a practical solution for increasing the output of RES-based systems such as large solar parks and wind farms. Around 5000 km of hydrogen pipelines are present in the world, compared to 2.91 million km of natural gas pipelines. The global natural gas network has extensive transport and storage capacities (>100 GWh), which is why lots of studies are investigating the possibilities of injecting hydrogen into existing infrastructure [77]. From there, it can be separated at the user site and used in different applications such as power generation, heat provision, urban buses, passenger cars, or other transport applications. Injecting green hydrogen into the gas grid offers a lot of advantages, such as sector coupling, gas decarbonization, energy storage, and easy distribution with the existing infrastructure. Furthermore, using the existing infrastructure presents significantly lower investment compared to the construction of new hydrogen pipelines, which is especially important in this development stage of the energy transition. However, various modifications of the system network, investigations of possible risks, safety measures, purification, blending, and extraction of hydrogen are all additional costs that have to be made. If more than 10 vol.% of hydrogen should be injected, the measuring instruments, control stations, and compressors have to be changed [78]. The maximum hydrogen blend level in the natural gas grid is usually 5–20 vol.%, potentially even 25% depending on the gas grid infrastructure. For example, in Belgium and the UK, maximum hydrogen levels are up to 3 vol.% (and the last projects aim at 20 vol.%), while in the Netherlands and Germany, they are up to 10–14 vol.% [79].

As discussed in Sect. 3, the current conventional technology used for hydrogen separation is PSA. In PSA systems, the separated and purified hydrogen is delivered at high pressure while the other gases are discharged at lower pressure. However, if the incoming gas mixture is from a high-pressure natural gas grid stream, the separated gases from the hydrogen need to be compressed to be sent back to the natural gas grid. For this reason, two mechanical compressors are required in the system. The first compressor is to reach the adsorption pressure for separation of hydrogen while the second one is necessary for compressing the natural gas back to the grid. Also, if such a system is employed for the separation of hydrogen in relatively low concentrations from a hydrogen-methane mixture, it would require a substantial amount of compression energy and capital for the reinjection of depleted hydrogen back to the grid. This makes PSA technology uneconomical. The recent NREL study showed the extraction cost for separating hydrogen from a 10% H_2 stream with a recovery of 80% is around $3.3–8.3/kg [80]. The PSA system works efficiently at large scale and higher hydrogen concentrations, but if the hydrogen in the stream is lower than 10%, the gas needs to be pressurized to a high-pressure ratio, which makes the PSA option inefficient [79]. For this reason, new systems for low hydrogen concentration separation from the natural gas grid, such as EHC systems, are necessary. EHC has the potential to efficiently and at a lower cost separate and purify hydrogen from the natural gas grid.

3.2.4 Hydrogen Production

Steam methane reforming (SMR) from natural gas, followed by hydrogen purification, currently supplies 76% of the world's hydrogen demand, totaling ~70 million tons per year. After SMR production, PSA is the most common choice for hydrogen purification. SMR implies the reaction between purified natural gas and super-heated steam in a high-temperature and high-pressure reformer furnace, producing mainly CO, H_2O, and H_2 [80]. Furthermore, over 90% of hydrogen comes from fossil fuels and industrial by-products. The hydrogen that comes from bioprocesses such as biomass pyrolysis and microbial fermentation contains impurities or by-products such as CO, CO_2, N_2, etc. from the raw materials or reaction process. Therefore, no matter how hydrogen is obtained, separation and the high level of hydrogen purity that can be achieved via EHC present a necessary and useful process [81]. The level of hydrogen purification after production depends on its afterward application, e.g., PEMFC requires high-purity hydrogen, while solid oxide fuel cells (SOFC) or hydrogen utilization in gas turbines for power generation require relatively low hydrogen purity. Regarding green hydrogen production via water electrolysis, EHC can be used at the outlet of an electrolyzer to compress produced hydrogen at very high pressure as it requires lower energy consumption at a lower pressure range, thus enhancing the efficiency of the whole system [7].

3.2.5 Hydrogen Application in the Transport Sector

The green energy transition progress is perhaps publicly most recognizable in the transport sector, where hydrogen is beginning to take a big role. Fuel cell and battery vehicles are both needed as replacements for fossil fuel-fueled vehicles to enable clean and green road transportation. Fuel cell electric vehicles (FCEVs) use hydrogen in PEMFC, which converts hydrogen from high-pressure storage vessels into electrical energy for the electric motor. When comparing FCEVs with a battery electric vehicle (BEV), a BEV with a range of about 500 km contains a lithium-ion battery energy storage system that has a mass of 830 kg, while FCEVs of the same range have hydrogen storage compressed to 700 bar that has a significantly lower amount of mass of 125 kg [82]. Furthermore, refueling hydrogen vessels takes a similar time as refueling fossil fuels, whereas charging batteries takes longer to allow time for the heat to dissipate from the battery. Therefore, vehicles that operate long distances for extended periods of the day (e.g., heavy-duty vehicles, public transportation, trucks, taxis) are more suitable for hydrogen that meets the range and operational time requirements. A conventional internal combustion gasoline engine is less than 20% efficient in converting the chemical energy in gasoline into power that moves the vehicle, under normal driving conditions, while FCEV with electric motor use 40–60% of the fuel's energy, which corresponds to more than a 50% reduction in fuel consumption compared to a conventional vehicle with a gasoline internal combustion engine [83].

Perhaps the biggest increase in the number of hydrogen-related transportation infrastructure and vehicles was at the end of 2019. According to data from the *IEA Advanced Fuel Cells Technology Collaboration Programme* [84], 25,212 FCEVs were in circulation in the world, where most of these vehicles are located in the United States of America, followed by China, South Korea, and Japan. 59% of them are located in Asia, 32% in North America, and 9% in Europe. The total number includes passenger and commercial vehicles, with the majority being passenger cars (75%), followed by buses (17.7%) delivery vehicles, and light trucks (7.3%). The European market is dominated by passenger cars from Germany and France with almost 50%, while Great Britain, The Netherlands, and Norway each have about 200 personal FCEVs. The number of all FCEVs in 2019 increased by 95%, while the total number of hydrogen refueling stations (HRSs) increased to 470, which represents an increase of 23% from the previous year.

In the process of electrochemical energy conversion under catalysis, hydrogen purity will directly affect the PEMFC life operation. Therefore, high-purity hydrogen is required for PEMFCs in FCEVs (>99.97%) (Society of Automotive Engineers International (SAE), SAE J2719:2020 Hydrogen Fuel Quality for Fuel Cell Vehicles; ISO 14687: 2019 Hydrogen quality requirements for PEM fuel cell road vehicle application) and hydrogen mobility (e.g., Hydrogen Mobility Europe (H2ME)) [26]. The standards provide background information and a hydrogen quality standard that HRSs need to provide for commercial PEMFC vehicles. The other requirement regarding HRSs is that for hydrogen storage, compression is needed due to hydrogen's low volumetric energy density as, e.g., 560 L of hydrogen compressed to 350 bar is an energy equivalent to 340 L of hydrogen compressed to 700 bar and 50 L of gasoline [85]. Therefore, given that high hydrogen pressures from 350 to 700 bar alongside high hydrogen purity are required for HRSs, the use of EHC is shown to be a good and efficient solution for HRS applications. Moreover, EHCs are low maintenance, vibration-free, quiet (suitable for operation in residential areas), have a long system lifetime, compact and reliable in comparison to multi-stage mechanical compressors that are currently the number one choice for HRS applications. Also, due to EHC scalability, it presents no problem in expanding HRS compression capacity with demand [75].

3.2.6 Hydrogen Storage and Distribution

In recent years, significant attention has been paid to the efficient use of hydrogen in a "hydrogen economy" that presents a potential way to deliver sustainable energy through the use of hydrogen. In this context, after being produced and before using it, hydrogen is stored and distributed. The most complex issues that require solutions due to hydrogen's physical properties are especially related to the latter two steps. As already explained in Sect. 1, because of its low specific volumetric energy density at atmospheric pressure, pressurized vessels for hydrogen storage and distribution are necessary [81]. The applications of hydrogen storage can be stationary or mobile. Stationary storage methods are mainly for on-site storage at either point of production or use, and stationary power generation. Mobile applications are usually for transporting the stored hydrogen to the point of storage or use, or hydrogen usage in FCEVs. Hydrogen distribution is a critical contributor to the cost, energy consumption, and GHG emissions associated with hydrogen pathways. The choice of the delivery method depends on specific geographic and market characteristics like target population and consuming behaviors, population density, size of HRS, and market penetration of FCEVs and other hydrogen-consuming systems [86]. Hydrogen's low specific volumetric energy density also makes it expensive to transport via road or ship. It has been shown that the cheapest hydrogen storage-delivery mode is obtained by compression and truck delivery, especially for small stations and low demands [7]. For this application, EHC could be used as a compression solution to fill hydrogen high-pressure vessels for storage and delivery. As hydrogen volume and required distance increase, trucks become a less feasible option and can be replaced by pipelines of compressed hydrogen, which can potentially transport thousands of tons per day [87].

3.2.7 Other Applications

Another application that EHCs could also be used for is recirculating hydrogen in a fuel cell stack. The recirculation of hydrogen via EHC on the anode side of a fuel cell can enable the voltage fluctuation and the number of purging operations to be decreased, avoiding additional pipe and resulting in a higher fuel cell efficiency than in the case of dead-ended mode [88]. In the same way, EHC could be used for recycling exhaust gas in industrial processes (e.g., semi-conductor industries) to increase sustainability as EHC can easily adapt to different exhaust flows and separate hydrogen even at low concentrations [65]. Also, as EHC ensures vibration-free operation with no moving parts, it could be ideal for the development of cryocoolers for infrared detectors in aerospace applications. They could replace the mechanical compressors, which are commonly used, because they produce significant vibration on spacecraft, causing substantial image artifacts [89]. In terms of hydrogen separation, EHC could be used to produce clean hydrogen used as a coolant in electric turbines, or for the cooling application in nuclear reactors (separation and concentration of hydrogen isotopes) [90].

3.3 The Safety Aspect of EHC Operation

The EHC system should be placed in a pressure-resistant container to comply with safety regulations regarding high-pressure systems. In the case of a hydrogen leakage, the nitrogen gas stream can be guided inside the container with a hydrogen sensor connected downstream in series to detect any hydrogen gas that leaks from the EHC cell and to purge the device of oxygen/air [91]. Leakage of a flammable or detonable mixture presents a crucial problem that can lead to a fire or detonation. When ignited, gas leakage can lead to deflagration and detonation hazards. The amplitude of the hazard depends on various factors such as the amount of hydrogen released, the air conditions, space limitation, and the location and strength of the ignition source. However, the risk is significantly lower in open space areas as hydrogen rapidly rises and disperses before it can ignite. Leakage is a much bigger problem in enclosed, unventilated spaces, where gas cannot easily escape from, which leads to high over-pressures. The so-called final protection measure for enclosed spaces to reduce the consequences of deflagrations in confined systems is explosion venting (e.g., pressure relief panels), where the vent size affects vented hydrogen/air explosion. Also, research conducted on hydrogen/air flammable gas cloud formation due to hydrogen leaks into confined spaces has shown that the speed of the gas release at a constant flow rate influences the diverse gas-dynamic patterns.

The low speed of hydrogen outflow initially forms a thin layer of flammable cloud at the ceiling, which then expands via concentration front downward, while the hydrogen high-speed results in uniform flammable cloud formation throughout the whole volume above the discharge point. Various initiatives (e.g., European Hydrogen Safety Panel—EHSP) and projects (e.g., Hydrogen Safety for Energy Applications—HySEA) are developing recommendations based on performative research, to improve international standards for hydrogen safety procedures such as the design of explosion venting devices [92].

When taking into account hydrogen properties in high-pressure gas storage, some aspects related to safety need to be considered. The low density of hydrogen, in conjunction with its small particle size, allows the penetration of hydrogen molecules in some metals and alloys such as cast iron and high-carbon steel. This penetration may end in small hydrogen leaks, or the presence of cracks within the wall and, therefore, it can contribute to the cracks spread, the decrease of the material strength, and the subsequent fracture. Furthermore, fatigue crack growth rates generally increase as hydrogen gas pressure increases. The main storage hazards are penetration in metals and alloys that can lead to embrittlement and explosion of hydrogen tanks when exposed to big temperature rise [93]. Hence, a good selection of materials is of preeminent importance. For example, the industry standard for components in hydrogen service is grade 316 stainless steels. Type 316/316 L is relatively immune to hydrogen embrittlement when exposed to high-pressure hydrogen and is preferred over other grades such as Type 304 L and Type 321. Also, copper-nickel (CuNi) is suitable for hydrogen service, and copper can be used for low-pressure applications [94]. The precautions for embrittlement avoidance in gas tanks are as follows:

1. Coat or surface treat of storage material to minimize the hydrogen penetration into metal materials.
2. Selection of the right material, mechanical loading, and pressure. For example, avoid high carbon steel, and use instead a higher alloyed steel with Nelson curves [95].

To prevent explosions in hydrogen tanks, the most common equipment often required by code implies a pressure release device (PRD) to prevent the overpressurization of gas vessels and distribution equipment. The PRD is intended for gaseous hydrogen operation, where the two triggers (pressure and temperature) activate the PRD and release the system. The PRD must be set to release the hydrogen flow over the maximum allowable working pressure of the system to prevent overpressure [96].

In case the accident does occur, risks from fire hazards could include personnel injuries, facility, and equipment damage due to high air temperatures, radiant heat fluxes, or contact with hydrogen flames, while explosion consequences encompass blast wave overpressure effects, the heat effects from subsequent fireballs, the impact from fragments, or the structure collapse. Named hydrogen risks and hazards can be compared to the ones associated with conventional fuel sources (gasoline and natural gas). To ensure safe operation and detect a leakage on time, reliable safety sensors should be used. National Fire Protection Association (NFPA) obligates the use of hydrogen detectors for indoor fueling operations [97]. The most used devices for fire prevention are monitoring systems, which are used as safety protection to detect hydrogen or its invisible flame. Examples of monitoring systems are ultraviolet (UV) flame detectors which sense the presence of UV radiation from unseen flames, LEL detection system that monitors for combustible gases, flare systems that expel flammable gases to the flare header, CO monitors (portable ones as well), and in some cases H_2S monitors as well. Safety measures to prevent the ignition of hydrogen include purging equipment with an inert gas feeding. If the oxygen in a whole system is not removed, a flammable mixture may still form. The purge effectiveness must be verified in compliance with applicable codes and regulations. Furthermore, it is recommended to use flammable gas detectors indoors, where hydrogen gas can be released or even accumulated, as well as a portable gas detector when performing maintenance. It is also important to eliminate the sources of ignition, such as sparks, open flames, smoking (e.g., cigarettes), and extremely hot objects from areas where hydrogen is processed, stored, or used. For this purpose, it is important to put up a warning sign indicating it, as well as to prescribe safety zones. This includes properly placing hydrogen systems at an appropriate distance from the exposures (e.g., buildings, sidewalks, vehicle parking areas, air intakes, power lines, etc.) and other hazardous materials following national codes and local regulations. In the EU each entity must perform a Risk Assessment of what to do under each circumstance. Integrated use of Failure Mode and Effect Analysis (FMEA), HAZard OPerability Analysis (HAZOP), and Fault Tree Analysis (FTA) are strongly advised to minimize risks. FMEA lowers the possibility of deficiencies, identifying the areas that are more prone to failures. In the case the failure occurs, FMEA minimizes its effects, applying the appropriate corrective actions. HAZOP analysis helps with the identification of incidents and potential scenarios, whereas FTA analysis evaluates their probability of happening [98]. Including all these well-established safety practices, alongside educated and trained personnel in hydrogen safety procedures, should be a standard in hydrogen safety engineering.

4 Outline

Hydrogen has higher specific energy than conventional fuels but compared per unit volume under normal conditions, its energy density is lower. Compressing hydrogen increases its density and allows for more efficient storage and transportation. Furthermore, to transfer hydrogen efficiently and at a low cost, it can be blended into the existing natural gas grid and purified for delivery to end-users. Also, alongside other uses, hydrogen separation is needed for production from fossil fuels and bioprocesses, which presently supply over 76% of the world's hydrogen demand. In order to be used in FCEVs, hydrogen also needs to be compressed to high pressures and fulfill a purity standard of 99.97% for its usage in PEMFC. Therefore, hydrogen compression and purification are a crucial step in the production, storage, transportation, and application of hydrogen. It follows that the creation of a hydrogen economy urgently necessitates a suitable compressor and purificator that is efficient, modular, safe, and inexpensive. Due to these characteristics, out of all compression technologies, electrochemical hydrogen compressors (EHC) stand out alongside additional advantages such as noiseless and vibration-free operation, nonexistence of moving parts, and high efficiency (usually over 60%). The modularity aspect of EHC is achievable by connecting multiple cells in appropriate stack configurations to achieve higher hydrogen pressures or flow rates, depending on a series or parallel hydrogen flow connections, respectively. Hydrogen purification and its separation from gaseous mixtures in a one-step operation are further benefits that give electrochemical compression another advantage. Given the current state of the art of purification technologies, EHCs can be a preferable option thanks to high selectivity, sensitivity to catalyst deactivation (e.g., CO deactivation), higher tolerance to "active" impurities (e.g., CO and CO_2) at higher temperatures, and the fact that hydrogen flux can be controlled by the current. The downsides of the EHC technology include the requirement for additional water and thermal management, and difficulty in manufacturing and gasketing the cell assembly due to high achieved pressures, which result in hydrogen back-diffusion, membrane deformation, and possible leakage. However, research regarding advanced water management improvement, catalyst optimization, enhanced thermal management, and the use of novel membranes could limit the EHC disadvantages. To conclude, EHC could be the required energy-efficient, low-cost technology for hydrogen compression and purification, pushing forward the energy transition towards green solutions, but since the technology is not yet mature, much progress in development and research is still needed. Nonetheless, the steep incline trend of new research on electrochemical compression is quite promising.

References

1. Lanz W (2001) Hydrogen properties . https://www1.eere.energy.gov/hydrogenandfuelcells/tech_validation/pdfs/fcm01r0.pdf. Accessed 9 Nov 2023
2. Marciuš D, Kovač A, Firak M (2022) Electrochemical hydrogen compressor: recent progress and challenges. Int J Hydrog Energy 47:24179–24193. https://doi.org/10.1016/j.ijhydene.2022.04.134
3. International Renewable Energy Agency (IRENA) (2021) Green hydrogen supply: a guide to policy making. IRENA, Abu Dhabi
4. Barthelemy H, Weber M, Barbier F (2017) Hydrogen storage: recent improvements and industrial perspectives. Int J Hydrog Energy 42:7254–7262. https://doi.org/10.1016/j.ijhydene.2016.03.178
5. Makridis S (2016) Hydrogen storage and compression. In: Carriveau R, Ting DS-K (eds) Methane and hydrogen for energy storage. Institution of Engineering and Technology, pp 1–28. https://doi.org/10.1049/PBPO101E_ch1
6. Rivard E, Trudeau M, Zaghib K (2019) Hydrogen storage for mobility: a review. Materials 12:1973. https://doi.org/10.3390/ma12121973

7. Sdanghi G, Maranzana G, Celzard A, Fierro V (2019) Review of the current technologies and performances of hydrogen compression for stationary and automotive applications. Renew Sust Energ Rev 102:150–170. https://doi.org/10.1016/j.rser.2018.11.028
8. Jin Y, Guo Y, Zhang S, Jiang J, Peng X (2023) Study on the dynamic characteristics of the free piston in the ionic liquid compressor for hydrogen refuelling stations by the fluid-structure interaction modelling. Int J Hydrog Energy 48:25410–25422. https://doi.org/10.1016/j.ijhydene.2023.03.202
9. Orlova S, Mezeckis N, Vasudev VPK (2023) Compression of hydrogen gas for energy storage: a review. Latv J Phys Tech Sci 60:4–16. https://doi.org/10.2478/lpts-2023-0007
10. Chalkiadakis N, Stubos A, Stamatakis E, Zoulias E, Tsoutsos T (2023) A review on hydrogen compression methods for hydrogen refuelling stations. In: Sankır M, Sankir N (eds) Hydrogen electrical vehicles, 1st edn. Wiley, pp 47–73. https://doi.org/10.1002/9781394167555.ch3
11. Zhou H, Dong P, Zhu S, Li S, Zhao S, Wang Y (2021) Design and theoretical analysis of a liquid piston hydrogen compressor. J Energy Storage 41:102861. https://doi.org/10.1016/j.est.2021.102861
12. Imtiaz A, Othman MHD, Jilani A, Khan IU, Kamaludin R, Iqbal J et al (2022) Challenges, opportunities and future directions of membrane technology for natural gas purification: a critical review. Membranes 12:646. https://doi.org/10.3390/membranes12070646
13. Chen H, Cong TN, Yang W, Tan C, Li Y, Ding Y (2009) Progress in electrical energy storage system: a critical review. Prog Nat Sci 19:291–312. https://doi.org/10.1016/j.pnsc.2008.07.014
14. Yang Z, Zhang J, Kintner-Meyer MCW, Lu X, Choi D, Lemmon JP et al (2011) Electrochemical energy storage for green grid. Chem Rev 111:3577–3613. https://doi.org/10.1021/cr100290v
15. Lototskyy MV, Tolj I, Pickering L, Sita C, Barbir F, Yartys V (2017) The use of metal hydrides in fuel cell applications. Prog Nat Sci Mater Int 27:3–20. https://doi.org/10.1016/j.pnsc.2017.01.008
16. Lee J, Yoon S, Park H, Chinannai MF, Phan TT, Kim S-K et al (2023) Coupled mechanical and electrochemical modeling and simulations for electrochemical hydrogen compressors (EHC). Renew Energy 216:119071. https://doi.org/10.1016/j.renene.2023.119071
17. Xiao J, Peng Y, Bénard P, Chahine R (2016) Thermal effects on breakthrough curves of pressure swing adsorption for hydrogen purification. Int J Hydrog Energy 41:8236–8245. https://doi.org/10.1016/j.ijhydene.2015.11.126
18. Adhikari S, Fernando S (2006) Hydrogen membrane separation techniques. Ind Eng Chem Res 45:875–881. https://doi.org/10.1021/ie050644l
19. Aasadnia M, Mehrpooya M, Ghorbani B (2021) A novel integrated structure for hydrogen purification using the cryogenic method. J Clean Prod 278:123872. https://doi.org/10.1016/j.jclepro.2020.123872
20. Zornoza B, Casado C, Navajas A (2013) Chapter 11 – Advances in hydrogen separation and purification with membrane technology. In: Gandía LM, Arzamendi G, Diéguez PM (eds) Renewable hydrogen technologies. Elsevier, Amsterdam, pp 245–268. https://doi.org/10.1016/B978-0-444-56352-1.00011-8
21. Watanabe M, Uchida H, Ohkubo K, Igarashi H (2003) Hydrogen purification for fuel cells: selective oxidation of carbon monoxide on Pt–Fe/zeolite catalysts. Appl Catal B Environ 46: 595–600. https://doi.org/10.1016/S0926-3373(03)00322-9
22. Dawood F, Anda M, Shafiullah GM (2020) Hydrogen production for energy: an overview. Int J Hydrog Energy 45:3847–3869. https://doi.org/10.1016/j.ijhydene.2019.12.059
23. Liemberger W, Groß M, Miltner M, Harasek M (2017) Experimental analysis of membrane and pressure swing adsorption (PSA) for the hydrogen separation from natural gas. J Clean Prod 167:896–907. https://doi.org/10.1016/j.jclepro.2017.08.012
24. Lu GQ, Diniz da Costa JC, Duke M, Giessler S, Socolow R, Williams RH et al (2007) Inorganic membranes for hydrogen production and purification: a critical review and perspective. J Colloid Interface Sci 314:589–603. https://doi.org/10.1016/j.jcis.2007.05.067

25. Li P, Wang Z, Qiao Z, Liu Y, Cao X, Li W et al (2015) Recent developments in membranes for efficient hydrogen purification. J Membr Sci 495:130–168. https://doi.org/10.1016/j.memsci.2015.08.010
26. Vermaak L, Neomagus HWJP, Bessarabov DG (2021) Recent advances in membrane-based electrochemical hydrogen separation: a review. Membranes 11:127. https://doi.org/10.3390/membranes11020127
27. Sdanghi G, Maranzana G, Celzard A, Fierro V (2020) Towards non-mechanical hybrid hydrogen compression for decentralized hydrogen facilities. Energies 13:3145. https://doi.org/10.3390/en13123145
28. Pineda-Delgado JL, Menchaca-Rivera JA, Pérez-Robles JF, Aviles-Arellano LM, Chávez-Ramirez AU, Gutierrez B CK et al (2022) Energetic evaluations of an electrochemical hydrogen compressor. J Energy Storage 55:105675. https://doi.org/10.1016/j.est.2022.105675
29. Pineda-Delgado JL, Chávez-Ramirez AU, Gutierrez B CK, Rivas S, Marisela C-R, de Jesús Hernández-Cortes R et al (2022) Effect of relative humidity and temperature on the performance of an electrochemical hydrogen compressor. Appl Energy 311:118617. https://doi.org/10.1016/j.apenergy.2022.118617
30. Chouhan A, Bahar B, Prasad AK (2020) Effect of back-diffusion on the performance of an electrochemical hydrogen compressor. Int J Hydrog Energy 45:10991–10999. https://doi.org/10.1016/j.ijhydene.2020.02.048
31. Nordio M, Eguaras Barain M, Raymakers L, Van Sint Annaland M, Mulder M, Gallucci F (2020) Effect of CO2 on the performance of an electrochemical hydrogen compressor. Chem Eng J 392:123647. https://doi.org/10.1016/j.cej.2019.123647
32. Hao YM, Nakajima H, Inada A, Sasaki K, Ito K (2019) Overpotentials and reaction mechanism in electrochemical hydrogen pumps. Electrochim Acta 301:274–283. https://doi.org/10.1016/j.electacta.2019.01.108
33. Durmus GNB, Colpan CO, Devrim Y (2021) A review on the development of the electrochemical hydrogen compressors. J Power Sources 494:229743. https://doi.org/10.1016/j.jpowsour.2021.229743
34. Yang D, Tan Y, Li B, Ming P, Xiao Q, Zhang C (2022) A review of the transition region of membrane electrode assembly of proton exchange membrane fuel cells: design, degradation, and mitigation. Membranes 12:306. https://doi.org/10.3390/membranes12030306
35. Ghosh PC (2013) Influences of contact pressure on the performances of polymer electrolyte fuel cells. J Energy 2013:1–11. https://doi.org/10.1155/2013/571389
36. Kahraman H, Orhan MF (2017) Flow field bipolar plates in a proton exchange membrane fuel cell: analysis & modeling. Energy Convers Manag 133:363–384. https://doi.org/10.1016/j.enconman.2016.10.053
37. Chouhan A, Aryal UR, Bahar B, Prasad AK (2020) Analysis of an electrochemical compressor stack. Int J Hydrog Energy 45:31452–31465. https://doi.org/10.1016/j.ijhydene.2020.08.164
38. Samad S, Loh KS, Wong WY, Lee TK, Sunarso J, Chong ST et al (2018) Carbon and non-carbon support materials for platinum-based catalysts in fuel cells. Int J Hydrog Energy 43:7823–7854. https://doi.org/10.1016/j.ijhydene.2018.02.154
39. Trégaro M, Rhandi M, Druart F, Deseure J, Chatenet M (2020) Electrochemical hydrogen compression and purification versus competing technologies: part II. Chall Electrocatal Chin J Catal 41:770–782. https://doi.org/10.1016/S1872-2067(19)63438-8
40. Ibeh B, Gardner C, Ternan M (2007) Separation of hydrogen from a hydrogen/methane mixture using a PEM fuel cell. Int J Hydrog Energy 32:908–914. https://doi.org/10.1016/j.ijhydene.2006.11.017
41. Gomez YA, Oyarce A, Lindbergh G, Lagergren C (2018) Ammonia contamination of a proton exchange membrane fuel cell. J Electrochem Soc 165:F189–F197. https://doi.org/10.1149/2.0761803jes
42. Tingelöf T, Hedström L, Holmström N, Alvfors P, Lindbergh G (2008) The influence of CO2, CO and air bleed on the current distribution of a polymer electrolyte fuel cell. Int J Hydrog Energy 33:2064–2072. https://doi.org/10.1016/j.ijhydene.2008.02.002

43. Gu T, Lee W-K, Zee JWV (2005) Quantifying the 'reverse water gas shift' reaction inside a PEM fuel cell. Appl Catal B Environ 56:43–50. https://doi.org/10.1016/j.apcatb.2004.08.016
44. Giner J (1963) Electrochemical reduction of CO2 on platinum electrodes in acid solutions. Electrochim Acta 8:857–865. https://doi.org/10.1016/0013-4686(63)80054-7
45. Gardner CL, Ternan M (2007) Electrochemical separation of hydrogen from reformate using PEM fuel cell technology. J Power Sources 171:835–841. https://doi.org/10.1016/j.jpowsour.2007.06.020
46. Jackson C, Raymakers LFJM, Mulder MJJ, Kucernak ARJ (2020) Poison mitigation strategies for the use of impure hydrogen in electrochemical hydrogen pumps and fuel cells. J Power Sources 472:228476. https://doi.org/10.1016/j.jpowsour.2020.228476
47. Jackson C, Raymakers LFJM, Mulder MJJ, Kucernak ARJ (2020) Assessing electrocatalyst hydrogen activity and CO tolerance: comparison of performance obtained using the high mass transport 'floating electrode' technique and in electrochemical hydrogen pumps. Appl Catal B Environ 268:118734. https://doi.org/10.1016/j.apcatb.2020.118734
48. Murdoch K, Blanchard R, Mukerjee S, Stracensky T, Sharma M, Pavlicek R et al (2019) Closed loop hydrogen recovery enabled by electrochemical hydrogen separation. ICES, Boston, p 20
49. Dubau L, Maillard F (2016) Unveiling the crucial role of temperature on the stability of oxygen reduction reaction electrocatalysts. Electrochem Commun 63:65–69. https://doi.org/10.1016/j.elecom.2015.12.011
50. Cheng X, Shi Z, Glass N, Zhang L, Zhang J, Song D et al (2007) A review of PEM hydrogen fuel cell contamination: impacts, mechanisms, and mitigation. J Power Sources 165:739–756. https://doi.org/10.1016/j.jpowsour.2006.12.012
51. Bouwman P (2014) Electrochemical hydrogen compression (EHC) solutions for hydrogen infrastructure. Fuel Cells Bull 2014:12–16. https://doi.org/10.1016/S1464-2859(14)70149-X
52. Ramousse J, Lottin O, Didierjean S, Maillet D (2009) Heat sources in proton exchange membrane (PEM) fuel cells. J Power Sources 192:435–441. https://doi.org/10.1016/j.jpowsour.2009.03.038
53. Nordio M, Rizzi F, Manzolini G, Mulder M, Raymakers L, Van Sint Annaland M et al (2019) Experimental and modelling study of an electrochemical hydrogen compressor. Chem Eng J 369:432–442. https://doi.org/10.1016/j.cej.2019.03.106
54. Nguyen M-T, Grigoriev SA, Kalinnikov AA, Filippov AA, Millet P, Fateev VN (2011) Characterisation of a electrochemical hydrogen pump using electrochemical impedance spectroscopy. J Appl Electrochem 41:1033–1042. https://doi.org/10.1007/s10800-011-0341-9
55. Hao Y, Nakajima H, Yoshizumi H, Inada A, Sasaki K, Ito K (2016) Characterization of an electrochemical hydrogen pump with internal humidifier and dead-end anode channel. Int J Hydrog Energy 41:13879–13887. https://doi.org/10.1016/j.ijhydene.2016.05.160
56. Sdanghi G, Dillet J, Didierjean S, Fierro V, Maranzana G (2020) Feasibility of hydrogen compression in an electrochemical system: focus on water transport mechanisms. Fuel Cells 20:370–380. https://doi.org/10.1002/fuce.201900068
57. Zou J, Han N, Yan J, Feng Q, Wang Y, Zhao Z et al (2020) Electrochemical compression technologies for high-pressure hydrogen: current status. Chall Perspect Electrochem Energy Rev. https://doi.org/10.1007/s41918-020-00077-0
58. Lipp L (2016) Electrochemical hydrogen compressor. Techical report DOE-FCE003727 fuel cell energy, Inc., Torrington, CT (United States). https://doi.org/10.2172/1235441
59. Ijaodola OS, El-Hassan Z, Ogungbemi E, Khatib FN, Wilberforce T, Thompson J et al (2019) Energy efficiency improvements by investigating the water flooding management on proton exchange membrane fuel cell (PEMFC). Energy 179:246–267. https://doi.org/10.1016/j.energy.2019.04.074
60. Gallagher KG, Pivovar BS, Fuller TF (2008) Electro-osmosis and water uptake in polymer electrolytes in equilibrium with water vapor at low temperatures. J Electrochem Soc 156:B330–B338
61. Rahbari A, Brenkman J, Hens R, Ramdin M, van den Broeke LJP, Schoon R et al (2019) Solubility of water in Hydrogen at high pressures: a molecular simulation study. J Chem Eng Data 64:4103–4115. https://doi.org/10.1021/acs.jced.9b00513

62. Bouwman PJ (2016) Advances in electrochemical hydrogen compression and purification. ECS Trans 75:503–510. https://doi.org/10.1149/07514.0503
63. Bossel U, Eliasson B (n.d.) Energy and the Hydrogen economy. US DOE, EERE, Washington, DC
64. Aziz M, Aryal UR, Prasad AK (2022) Optimization of electrochemical hydrogen compression through computational modeling. Int J Hydrog Energy:S0360319922032694. https://doi.org/10.1016/j.ijhydene.2022.07.256
65. Rhandi M, Trégaro M, Druart F, Deseure J, Chatenet M (2020) Electrochemical hydrogen compression and purification versus competing technologies: part I. Pros and Cons Chin J Catal 41:756–769. https://doi.org/10.1016/S1872-2067(19)63404-2
66. Williams M, Song X, Gemmen R (2021) Research needs for the solid oxide electrolyzer (SOEC) with a proton-conducting SOEC (H-SOEC) electrochemical hydrogen compressor (EHC) energy conversion network (ECN). ECS Trans 103:451–474. https://doi.org/10.1149/10301.0451ecst
67. Bampaou M, Panopoulos KD, Papadopoulos AI, Seferlis P, Voutetakis S (2018) An electrochemical hydrogen compression model. Chem Eng Trans 70:1213–1218. https://doi.org/10.3303/CET1870203
68. Dale NV (2009) Characterization of PEM electrolyzer and PEM fuel cell stacks using electrochemical impedance spectroscopy. Dissertation. University of North Dakota
69. Springer TE, Zawodzinski TA, Gottesfeld S (1991) Polymer electrolyte fuel cell model. J Electrochem Soc 138:2334–2342. https://doi.org/10.1149/1.2085971
70. Zawodzinski TA, Davey J, Valerio J, Gottesfeld S (1995) The water content dependence of electro-osmotic drag in proton-conducting polymer electrolytes. Electrochim Acta 40:297–302. https://doi.org/10.1016/0013-4686(94)00277-8
71. Yodwong B, Guilbert D, Phattanasak M, Kaewmanee W, Hinaje M, Vitale G (2020) Faraday's efficiency modeling of a proton exchange membrane Electrolyzer based on experimental data. Energies 13:4792. https://doi.org/10.3390/en13184792
72. Casati C, Longhi P, Zanderighi L, Bianchi F (2008) Some fundamental aspects in electrochemical hydrogen purification/compression. J Power Sources 180:103–113. https://doi.org/10.1016/j.jpowsour.2008.01.096
73. Moton JM, James BD, Colella WG (2014) Advances in electrochemical compression of hydrogen. ASME 2014 12th international conference on fuel cell science, engineering and technology. American Society of Mechanical Engineers, Boston, p V001T01A002. https://doi.org/10.1115/FuelCell2014-6641
74. Zou J, Jin Y, Wen Z, Xing S, Han N, Yao K et al (2020) Insights into electrochemical hydrogen compressor operating parameters and membrane electrode assembly degradation mechanisms. J Power Sources:229249. https://doi.org/10.1016/j.jpowsour.2020.229249
75. HyET Hydrogen (2020) Compression in one single stage. https://hyethydrogen.com/compression-in-one-single-stage/. Accessed 7 Jan 2024
76. Hamdan M (2018) Report electrochemical compression 2018. Technical Report DE-EE0007647 DOE Hydrogen and Fuel Cells Program. Giner ELX 89 Rumford Ave. Newton, MA 02466. https://www.hydrogen.energy.gov/docs/hydrogenprogramlibraries/pdfs/review18/pd136_hamdan_2018_o.pdf
77. Kovač A, Paranos M, Marciuš D (2021) Hydrogen in energy transition: a review. Int J Hydrog Energy 46:10016–10035. https://doi.org/10.1016/j.ijhydene.2020.11.256
78. Deymi-Dashtebayaz M, Ebrahimi-Moghadam A, Pishbin SI, Pourramezan M (2019) Investigating the effect of hydrogen injection on natural gas thermo-physical properties with various compositions. Energy 167:235–245. https://doi.org/10.1016/j.energy.2018.10.186
79. Nordio M, Wassie SA, Van Sint Annaland M, Pacheco Tanaka DA, Viviente Sole JL, Gallucci F (2021) Techno-economic evaluation on a hybrid technology for low hydrogen concentration separation and purification from natural gas grid. Int J Hydrog Energy 46:23417–23435. https://doi.org/10.1016/j.ijhydene.2020.05.009

80. Sweeney DM, Alves V, Sakhai S, Dinh S, Lima FV (2023) Techno-economic analysis and optimization of intensified, large-scale hydrogen production with membrane reactors. Ind Eng Chem Res 62:19740–19751. https://doi.org/10.1021/acs.iecr.3c02045
81. Cheng A, Xiao W, Jiang X, Ruan X, He G, Li X et al (2023) A partial element stage cut electrochemical hydrogen pump model for hydrogen separation and compression. Sep Purif Technol 307:122790. https://doi.org/10.1016/j.seppur.2022.122790
82. Ajanovic A, Haas R (2021) Prospects and impediments for hydrogen and fuel cell vehicles in the transport sector. Int J Hydrog Energy 46:10049–10058. https://doi.org/10.1016/j.ijhydene.2020.03.122
83. Dash SK, Chakraborty S, Roccotelli M, Sahu UK (2022) Hydrogen fuel for future mobility: challenges and future aspects. Sustain For 14:8285. https://doi.org/10.3390/su14148285
84. Can Samsun R, Antoni L, Rex M, Stolten D (2021) Deployment status of fuel cells in road transport: 2021 update. Energy Environ 542:5–11
85. Tzimas E (2003) In: Europäische Kommission (ed) Hydrogen storage: state-of-the-art and future perspective. Office for Official Publications of the European Communities, Luxembourg
86. Moradi R, Groth KM (2019) Hydrogen storage and delivery: review of the state of the art technologies and risk and reliability analysis. Int J Hydrog Energy 44:12254–12269. https://doi.org/10.1016/j.ijhydene.2019.03.041
87. Bloomberg NEF (2020) Hydrogen economy outlook 2020. Bloomberg Finance L.P, New York
88. Barbir F, Görgün H (2007) Electrochemical hydrogen pump for recirculation of hydrogen in a fuel cell stack. J Appl Electrochem 37:359–365. https://doi.org/10.1007/s10800-006-9266-0
89. Chen W, Smith B, Zagarola M, Narayanan S (2012) A vibration-free 35 K Cryocooler driven by electrochemical compressors. International Cryocooler Conference, Inc, Boulder, p 10
90. Grigoriev SA, Shtatniy IG, Millet P, Porembsky VI, Fateev VN (2011) Description and characterization of an electrochemical hydrogen compressor/concentrator based on solid polymer electrolyte technology. Int J Hydrog Energy 36:4148–4155. https://doi.org/10.1016/j.ijhydene.2010.07.012
91. Ito K, Hao YM, Nakajima H, Yoshizumi H, Sasaki K (2013) Electrochemical characterization of hydrogen pump with internal humidifier and dead-end anode channel. ECS Trans 58:681–691. https://doi.org/10.1149/05801.0681ecst
92. Skjold T, Hisken H, Lakshmipathy S, Atanga G, Carcassi M, Schiavetti M et al (2019) Blind-prediction: estimating the consequences of vented hydrogen deflagrations for homogeneous mixtures in 20-foot ISO containers. Int J Hydrog Energy 44:8997–9008. https://doi.org/10.1016/j.ijhydene.2018.06.191
93. Louthan MR (2008) Hydrogen embrittlement of metals: a primer for the failure analyst. J Fail Anal Prev 8:289–307. https://doi.org/10.1007/s11668-008-9133-x
94. Dwivedi SK, Vishwakarma M (2018) Hydrogen embrittlement in different materials: A review. Int J Hydrog Energy 43:21603–21616. https://doi.org/10.1016/j.ijhydene.2018.09.201
95. Poorhaydari K (2021) A comprehensive examination of high-temperature hydrogen attack—a review of over a century of investigations. J Mater Eng Perform 30:7875–7908. https://doi.org/10.1007/s11665-021-06045-z
96. Kostival A, Rivkin C, Buttner W, Burgess R (2013) Pressure relief devices for high-pressure gaseous storage systems: applicability to hydrogen technology. https://doi.org/10.2172/1110463
97. Buttner WJ, Post MB, Burgess R, Rivkin C (2011) An overview of hydrogen safety sensors and requirements. Int J Hydrog Energy 36:2462–2470. https://doi.org/10.1016/j.ijhydene.2010.04.176
98. Casamirra M, Castiglia F, Giardina M, Lombardo C (2009) Safety studies of a hydrogen refuelling station: determination of the occurrence frequency of the accidental scenarios. Int J Hydrog Energy 34:5846–5854. https://doi.org/10.1016/j.ijhydene.2009.01.096

The Role of Hydrogen in Energy System: State of Art and Future Prospects

Amela Ajanovic and Marlene Sayer

1 Introduction

The energy system, which encompasses the complex network of infrastructure, technologies and resources needed for the production, conversion, distribution, and providing energy services, is going through a huge transformation. Due to environmental problems, transition from fossil fuels toward renewables and low-carbon alternative fuels is needed, and it is widely supported all over the world. Over the last decade, there has been a significant increase in the use of renewable energy sources, especially wind and solar energy, worldwide (see Fig. 1). Currently, about one-seventh of the world's primary energy is sourced from renewable technologies [1]. This positive development, which can contribute to the reduction of the energy import dependency and increase of energy supply security, as well as to the reduction of the global greenhouse gas (GHG) emissions, is also creating some new challenges. Since wind and solar energy are variable energy sources that are not always available according to our needs, the rising question is what to do with the surplus electricity from renewable energy sources, and how to provide energy in the time periods in which electricity demand is higher than supply. To integrate a higher share of renewable energy sources in the electricity grid, what is the crucial aspect of the transformation of the energy system? It will need advancements in grid management, better demand-side management, and increasing use of different storage options.

Energy storage technologies, including batteries, pumped hydro, and other storage methods, play a crucial role in balancing supply and demand, especially with the increased use of intermittent renewable energy sources. In the future, storage technologies will contribute even more to a reliable and resilient energy system.

A. Ajanovic (✉) · M. Sayer
Energy Economics Group, Vienna University of Technology—TU WIEN, Vienna, Austria
e-mail: ajanovic@eeg.tuwien.ac.at; sayer@eeg.tuwien.ac.at

A. Kovač (ed.), *Challenges and Solutions in the Hydrogen Value Chain*,
https://doi.org/10.1007/978-3-031-69051-8_5

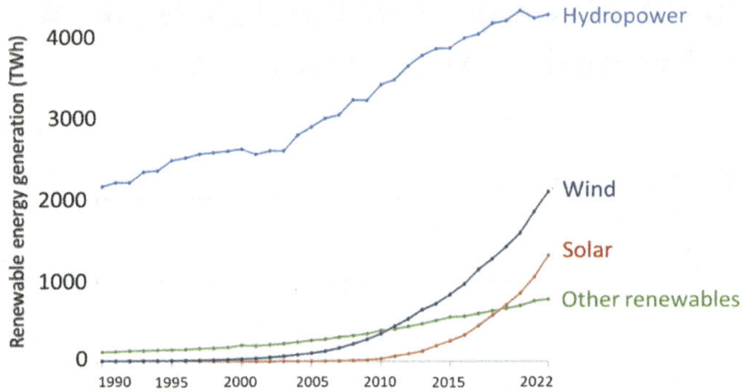

Fig. 1 Development of the global renewable energy generation by source [1]

In this context, hydrogen could play a significant role in the future. Hydrogen can be used to store excess renewable electricity generated during periods in which electricity production is higher than electricity demand, and then, this hydrogen can be used for electricity generation during times with no or low availability of renewable energy, contributing to the grid balance.

Actually, hydrogen, as a versatile and clean energy carrier, can play different roles in the energy system. Hydrogen is like electricity, a secondary energy carrier that can be produced from different primary energy sources using various methods. It can be gasified or liquified, stored and transported, and finally used in different sectors, such as transportation, industry, and power generation. Currently, hydrogen is mostly used in industry sector. However, with the increasing need for the decarbonization of all sectors, the role of hydrogen is becoming more important. A replacement of conventional fossil fuels with hydrogen could significantly reduce GHG emissions. In this context, of special interest is clean hydrogen, which is produced using renewable energy sources. Clean hydrogen can play an important role in the transformation of the energy system. The transformation is a complex process that involves all sectors and requires collaboration among governments, businesses, communities, and individuals.

The idea to use hydrogen in the energy system is actually very old. Already in 1874, the first vision of a hydrogen economy was presented by Jules Verne in his work "The Mysterious Island." Also, the concept of renewable hydrogen was presented in 1920 by British scientist J.B.S. Haldane in his paper Science and the Future by proposing that "there will be great power stations where during windy weather the surplus power will be used for the electrolytic composition of water into oxygen and hydrogen." Around the same time, hydrogen started being employed in industrial processes, particularly for ammonia synthesis. Additionally, until the 1960s, many cities utilized town gas, containing 50% hydrogen, for illuminating streets and powering households [2]. In 1958, NASA integrated hydrogen into its space program, primarily for rocket propulsion and fuel cells. Simultaneously, research was underway for other applications of hydrogen, leading to the development of the first PEM fuel cell. The term hydrogen economy emerged in the 1970s, initially used by General Motors to advocate for hydrogen use in the transportation sector [2]. Overall, the renewed interest in hydrogen, beyond its industrial applications, can be attributed to the price increase of oil during the oil crisis and increased environmental awareness during that time. In the mid-1980s, the German Center for Aeronautics and Space (DLR) further elaborated on the concept of a hydrogen economy in the energy sector, proposing a practical approach that involves utilizing various primary energy sources to generate electricity, followed by hydrogen production through electrolysis [3].

The appeal for hydrogen resurfaced in the early twenty-first century, driven by technological advancements in the 1990s (e.g., the introduction of the first proton-exchange membrane fuel cell (PEMFC) car in 1993). Rifkin (2002) [4] discussed hydrogen as an environmentally friendly universal energy carrier, envisioning a shift toward a sustainable energy system. In his influential work, "The Hydrogen Economy," Rifkin anticipates the dawn of a new economy powered by hydrogen, fundamentally altering market, political, and social institutions reminiscent of the transition to coal and steam power at the start of the industrial age.

Additional research explicitly addressing a hydrogen-centric future energy system includes works by Ball and Weeda (2015) [2], Rosen and Koohi-Fayegh (2016) [5], and Capurso et al. (2022) [6]. The central theme in these studies is the pivotal role of green hydrogen, highlighting the necessity of low electricity costs sourced from renewables. Abdin et al. (2020) [7] attribute the renewed interest in a hydrogen economy to a significant reduction in the cost of renewable energies, impacting politics, media, and the scientific community. The question of whether this resurgence will lead to tangible implementations in the coming decades or remain a temporary trend, as observed in the past, remains uncertain [8]. Nevertheless, the consensus across most studies is that both electricity and hydrogen will play vital roles as future energy carriers.

In the work focusing on the U.S. economy, Majumdar et al. (2021) [9] advocate to formulate a hydrogen strategy, invest in research and development, implement hydrogen policies, and integrate Carbon Capture, Utilization, and Storage (CCUS) technologies for the future realization of a hydrogen economy. Mac Dowell et al. [10] stress the importance of defining zero-transition pathways in accordance with socio-political-economic constraints. The need to consider socio-geopolitical implications and establish consistent international regulations to prevent potential green colonialism in the future hydrogen-based economy is highlighted by Akhtar et al. (2023) [11], a concern also acknowledged by Dillman and Heinonen [12].

Several studies focus on the practicalities of implementing a hydrogen economy in specific countries, such as [13] for India, [14] for Australia, and [15] for Austria. Schlund et al. (2022) [16] identify key stakeholders in Germany responsible for driving the expansion of hydrogen usage, pointing out electricity utilities, research and development entities, and hydrogen technology providers as the most influential groups.

Over the last few decades, the research on hydrogen has been intensified, however, with many ups and downs. Although hydrogen has many potential benefits, there are some major challenges such as hydrogen cost, efficiency of hydrogen production methods as well as whole supply chain, the investment and development of the infrastructure for hydrogen distribution and storage.

However, in spite of all the challenges and unfulfilled expectations in the past, hydrogen is still seen as a key player in the transition to a more sustainable, decarbonized energy system. Its versatility makes it attractive for the use in various applications across different sectors.

Major goal of this paper is to document state of the art in the production and use of hydrogen, with the focus on green hydrogen and possible new applications of hydrogen. Of special interest is hydrogen use in the transport sector, especially in the cases in which electrification is not an appropriate solution.

In this chapter, at first different hydrogen production methods are documented and discussed. As next, current and future hydrogen applications are analyzed with the focus on hydrogen use in the transport sector. Major conclusions and outlook are presented at the end of the work.

2 Hydrogen Production

There are different methods for hydrogen production. Basically, they can be split into three categories: thermal, electrochemical, and biological methods. In these hydrogen-production methods, different feedstocks, such as water, natural gas, coal, biomass, and different energy sources, could be used resulting in different emissions from hydrogen production. All biological methods for hydrogen production are still in the early stages of the development. Currently, the most commonly used are thermal methods, especially steam reforming process.

The choice of production method depends on different factors such as cost, availability of feedstocks and energy sources, existing infrastructure, purpose of the final use, and environmental goals.

The methods of hydrogen production are provided in the Table 1.

In spite of the variety of the methods for hydrogen production, hydrogen is still mostly produced by steam methane reforming. As shown in Fig. 2, current hydrogen production is largely based on fossil energy, mostly natural gas and coal. From 95 Mt. of hydrogen produced globally in 2022, less than 1 Mt. can be considered low-emission hydrogen [20]. Although the number of electrolysis projects is significantly increasing, water electrolysis contributed with less than 100 kt hydrogen in 2022 [20].

Depending on the selected hydrogen production method as well as the choice of primary energy sources used for hydrogen production, hydrogen generation can be classified based on different colors. In major classification three colors are used: green—for hydrogen produced using renewable energy sources, gray—for hydrogen produced from fossil energy, and blue—for hydrogen production from fossil energy in combination with carbon capture, storage, or use. However, in literature also, other colors are used to describe more precisely which kind of primary source or

Table 1 The various hydrogen production technologies [17–19]

Hydrogen production technologies	Short description
Steam methane reforming (SMR)	Process: SMR is the most common method for producing hydrogen and involves the reaction of methane with steam in the presence of a catalyst to produce hydrogen and carbon dioxide Feedstock: Natural gas is the primary feedstock, and the process is sometimes referred to as "gray hydrogen" when the carbon emissions are not captured and stored Efficiency: 60% (small plant)—85% (large plant)
Electrolysis	Process: Electrolysis involves using an electric current to split water into hydrogen and oxygen. This can be done through two main methods: alkaline electrolysis and proton exchange membrane (PEM) electrolysis Feedstock: Water Energy source: Electricity Efficiency: Alkaline: 62% (small plant)—82% (large plant); PEM: 56% (small plant)—84% (large plant)
Partial oxidation of hydrocarbons	Process: Partial oxidation involves reacting a hydrocarbon feedstock, such as natural gas or liquid fuels, with a limited amount of oxygen. This process produces hydrogen, carbon monoxide, and in some cases, carbon dioxide Feedstock: Natural gas or liquid hydrocarbons Efficiency: 55% (small plant)—75% (large plant)
Biomass gasification	Process: Biomass gasification involves heating biomass materials (such as wood, agricultural residues, or organic waste) in a controlled environment to produce a mixture of hydrogen, carbon monoxide, and other gases Feedstock: Biomass Efficiency: 74% (small plant)—85% (large plant)
Coal gasification	Process: Coal gasification is a process that turns coal into a very hot gaseous mixture of carbon monoxide, hydrogen, carbon dioxide and other compounds. This process allows coal use in a more environmentally friendly way compared to traditional combustion Feedstock: Coal Efficiency: 35% (small plant)—50% (large plant)
Thermochemical water splitting	Process: Thermochemical processes use heat to break the chemical bonds in water molecules, releasing hydrogen and oxygen. High-temperature reactors are typically required for this method Energy source: Heat can be provided by various means, including solar, nuclear, or concentrated solar power Efficiency: 52% (900 °C)—60% (1000 °C)

technology is used. For example, the use of fossil energy for hydrogen production can be even more exactly described using the colors brown or black. In some papers, brown hydrogen is used for coal gasification [21, 22]. Sometimes, there is also a distinction between black and brown coal, resulting in brown hydrogen when brown coal is used and black hydrogen when black coal is used [23].

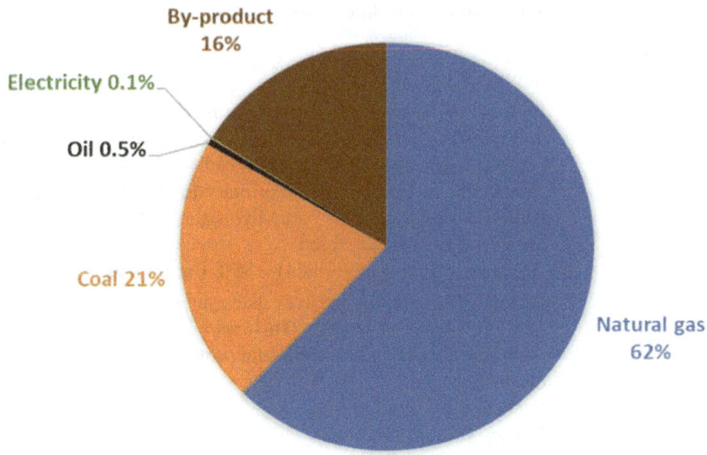

Fig. 2 Share of hydrogen production by energy used in the process. (Data source [20])

Table 2 Colors of hydrogen [24, 25]

Colors of hydrogen		Short description
Black H2	Brown H2	hydrogen extracted from black or brown coal using gasification
Grey H2		hydrogen produced by fossil fuels, mostly natural gas using steam methane reformation
Blue H2		hydrogen produced from fossil fuels in combination with carbon dioxide capture, storage or use.
Turquoise H2		hydrogen produced from fossil fuels via pyrolysis, where by-product is solid carbon
Green H2		hydrogen produced by electrolysis using electricity from renewable energy sources
Pink H2		hydrogen produced by electrolysis using electricity from nuclear power plants
Yellow H2		hydrogen produced by electrolysis using grid electricity from various sources
White H2		hydrogen produced as a by-product in industry, or hydrogen occurring in its natural form.

Comprehensive list of hydrogen colors is presented in Table 2. A more detailed description of different colors of hydrogen is provided by Ajanovic et al. (2022) [24].

Although the largest amount of hydrogen is currently produced by steam reforming of natural gas, in the future, the largest expectations are for green hydrogen, which should be produced using surplus electricity from variable renewable sources. Such kind of hydrogen is of special interest in the transformation of the

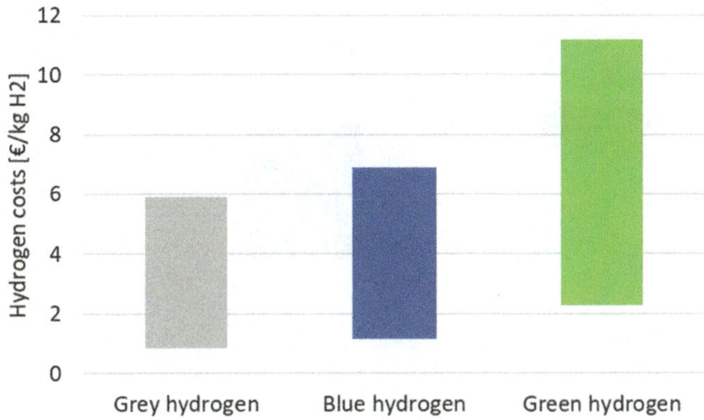

Fig. 3 Hydrogen production costs. (Data sources: [17, 20, 24, 31–35])

current energy system and transition toward more sustainable one. In literature, green hydrogen is sometimes also described using expressions such as clean hydrogen, renewable hydrogen, or low-carbon hydrogen.

Even if there are different electrolysis technologies, currently mostly used is alkaline water electrolysis. This is the most mature electrolyser technology, which has been in use since 1920 and has a market share of about 70% [26]. However, as in the case of every technology, alkaline water electrolysis has some advantages and disadvantages. Major advantages are the high maturity level in comparison to other electrolysis technologies [27, 28], low costs, long operational life, and suitability for large-scale applications [24]. A disadvantage is that this process needs to run continuously to avoid damage, which could be a problem if we want to use only surplus electricity from variable renewable energy sources. Alkaline water electrolysis tends to have lower efficiency compared to solid oxide electrolyser cells [29]. A comparative study of alkaline water electrolysis in comparison with proton exchange membrane water electrolysis and solid oxide electrolysis is provided by Hu et al. (2022) [30].

Globally, water electrolysis accounts currently for only about 0.1% of total hydrogen production [20]. One of the major reasons for this is the large cost difference between hydrogen production pathways. Currently, the production costs of gray hydrogen are the lowest. Due to this, gray hydrogen is the choice number one for the different industry applications in which economics has priority over environment. Figure 3 shows the range of hydrogen production costs for green, gray, and blue hydrogen based on literature [17, 20, 24, 31–35]. In the literature, there is a wide range of hydrogen production costs depending on the technology used as well as due to different assumptions regarding full load hours and energy prices.

The hydrogen production costs are dependent on the technology as well as on the costs of energy used. Although, hydrogen production by steam reforming is the cheapest process for hydrogen production, in 2022, due to the war in Ukraine and increasing natural gas prices, the production costs of gray hydrogen increased significantly.

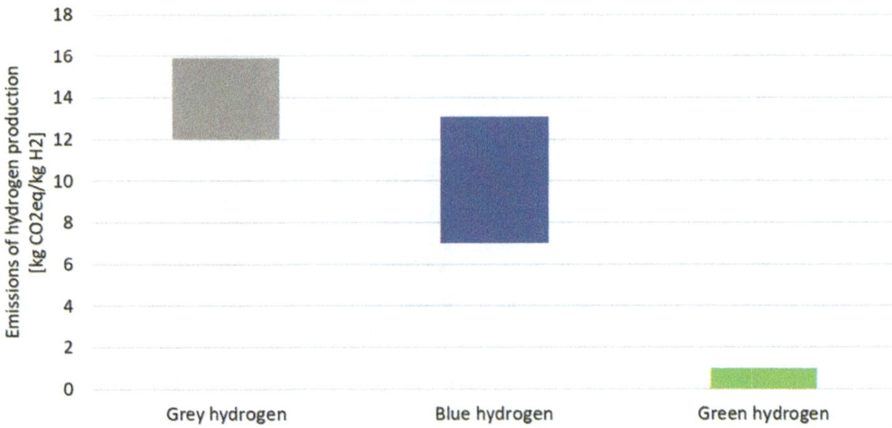

Fig. 4 Emissions of different hydrogen production technologies [20, 24]

Since the cost of hydrogen production by electrolysis is largely determined by the capital cost of electrolyser, of special interest are the number of full load hours and electricity prices. As calculated by Ajanovic et al. (2024), the lowest cost of green hydrogen could be reached at about 7000 full load hours per year [18].

Currently, due to the pressing environmental problems and policy goals set, as well as with increasing cost-effectiveness of renewable energy, the shift toward green hydrogen is on the rise. The emissions from hydrogen production are in a broad range, depending on the energy and technology used. In the case that natural gas is used, per kg of hydrogen, about 9 kg of CO_2 equivalent are produced [20]. In combination with carbon capture and storage, these emissions can be significantly reduced. However, as shown in Fig. 4, green hydrogen can contribute to the largest emission reduction. It is important to stress that the use of the current global grid electricity mix in electrolyser for hydrogen production (yellow hydrogen) results in the almost same amount of emissions as by gray hydrogen.

3 Current and Future Hydrogen Applications

Due to its versatility, hydrogen has a broad portfolio of possible applications in different sectors. In 2022, hydrogen use reached 95 Mt. worldwide [20]. The largest amount of this hydrogen was used in China (29%), North America (17%), Middle East (13%), India (9%), and Europe (8%).

Over the last years, it has a slight increase in the global hydrogen use. However, this increase was not driven by policies focusing on decarbonization and transformation of energy system. Currently, the largest amount of hydrogen is produced from fossil energy, so that increasing use of hydrogen is not contributing to the emission reduction. For instance, in 2022, hydrogen use in refining reached more

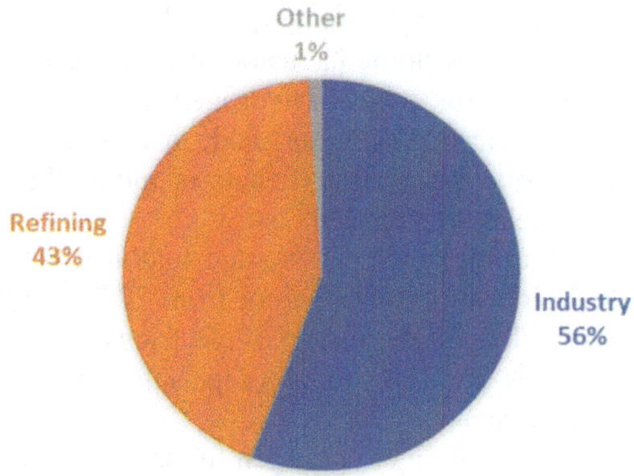

Fig. 5 Hydrogen use by sector [20]

than 41 Mt., but less than 1% of it was produced using low-emission technologies (see Fig. 5). Additional 53 Mt. of hydrogen were used in industry, however, almost all this hydrogen was produced from fossil fuels. Consequently, industrial hydrogen production was responsible for 680 Mt. of CO2 emissions in 2022 [20].

Traditionally, hydrogen is mostly used in petroleum refining processes, such as hydrocracking and desulfurization, to improve the quality of fuels, as well as in different industry processes. For example, in the chemical industry, hydrogen is used as a feedstock. It is a crucial feedstock in the production of ammonia, methanol, and other chemicals. In the steel industry, hydrogen is used as a reducing agent. In addition, hydrogen is also used in the food industry for the hydrogenation of oils, converting liquid oils into solid fats for various food products. From a European standpoint, the predominant use of hydrogen within the EU at present is in the chemical industry for the production of plastics and fertilizers [36].

All these applications are standard applications that were already established long time ago. In such applications, it was important to have cheap and easily produced hydrogen, and environmental aspects were more or less completely neglected. Over the last decade, due to the pressing environmental problems, the interest in decarbonization of already hydrogen-consuming applications as well as in development of new applications for the use of clean hydrogen has been rising.

Hydrogen substitution is frequently advocated for processes like oil refining and ammonia production. This seems quite uncomplicated, especially when one considers that electrical power from the grid can be utilized in production, given the constant energy requirement of the Haber Bosch process. The only drawback lies in the production of urea, the widely used fertilizer, which necessitates CO_2 inputs. In traditional methods, these are sourced from natural gas used in the SMR process, but in the case of green hydrogen, alternative sources must be sought [37]. Hydrogen's potential use in steelmaking is frequently debated as an alternative. However, careful consideration is imperative for all industrial applications where hydrogen isn't employed as feedstock, given the existence of various alternatives for direct electrification [38]. According to the study by Madeddu et al. (2020) [38], around 78% of European industrial heat could shift to decarbonization through clean power and currently available commercial technologies. This percentage could escalate to as high as 99% with the inclusion of technologies still in the developmental phase.

Newer applications of hydrogen span various sectors, spanning from transport to industry, the power sector, and, in niche areas, buildings for electricity and heat production. However, efficiencies along the chains up to end-use technologies remain relatively modest.

One example is the storage and re-electrification of hydrogen, with a total conversion efficiency ranging from 27% to 38%. Despite several necessary steps, seasonal/long-term storage of electricity currently relies on hydrogen underground storage as the sole option, especially to accommodate variable renewables and enhance capacity [39].

The utilization of hydrogen for space heating in buildings is largely dismissed in most literature. Its low maturity in comparison to alternatives like heat pumps, coupled with significantly lower efficiency, is a notable drawback [40]. In a comprehensive analysis, Rosenow (2022) [41] examines 32 "independent" studies (identified by the author as not influenced by specific industries such as gas, oil, electricity, heat pumps, or boiler manufacturers) and reveals that none of them ascribe a meaningful role to space heating in the Energy Transition. It is crucial to distinguish between independent papers and reports sponsored by the gas, gas grid, and heating industries in this context [42]. Even in the latest future scenarios outlined by Shell, hydrogen is projected to hold only a 0.4% share in the space heating sector [43].

A unique scenario involves the potential integration of hydrogen into the existing natural gas network. Here, the aim is to introduce green hydrogen volumes to mitigate overall gas consumption emissions. Nevertheless, a study conducted by the Fraunhofer Institute [44] contends that the technical complexities of this blending process (up to 20% of natural gas) are disproportionate to the potential greenhouse gas (GHG) savings, estimated to be between 6% and 7% owing to the lower calorific value of hydrogen in comparison to natural gas. This conclusion supports IRENA's technology review of hydrogen carriers report [45]. The Fraunhofer study further highlights the limited availability of green hydrogen as a problematic factor in the near future. The emphasis is on replacing applications with substantial CO_2 reduction potential with hydrogen rather than opting for limited emission reductions through blending, which directly competes with the direct use of hydrogen [44].

It is crucial to give priority to hydrogen in end-use applications where there are no viable alternatives. With respect to this, of special interest is hydrogen use in the transport sector, which is the least diversified sector. Current transport system is largely based on fossil fuels. Increasing electrification over the last decade has been widely supported and promoted. However, there are some challenges related to the use of battery electric vehicles, such as energy density, driving range, and charging time, so that electricity is not suitable for all using patterns and applications in all transport modes. Furthermore, in the transportation sector, where price levels are already elevated, there exists the potential for higher profit margins for hydrogen [46]. Sectors like steel and chemicals will demand substantial amounts of hydrogen, but they are also the most price-sensitive industries and face global competition [47].

4 Transport Sector

The idea to use hydrogen in the transport sector is also an old one. Already in 1807, a Swiss inventor, Francois Isaac de Rivaz, built one of the first automobiles with an engine powered by a mixture of hydrogen and oxygen [48].

Starting from the 1960s, there were many attempts to use hydrogen in the different transport modes, such as farm tractors, vans, cars, buses, forklifts, and tramcars (see Fig. 6). Moreover, hydrogen is used as rocket fuel for space exploration as well as in fuel cell-powered submarines that undergo deep-water trials.

Over the last decade, several automakers, such as Toyota, Hyundai, and Honda, are increasingly investing in hydrogen-powered vehicles. Toyota is a pioneer in fuel cell technology and has commercially released the Toyota Mirai, the first mass-produced hydrogen fuel cell vehicle.

In general, hydrogen can be used as a fuel in fuel cell vehicles (FCV), where it reacts with oxygen in a fuel cell to produce electricity, which is then used to power the vehicle's electric motor (see Fig. 7). FCVs convert hydrogen into electricity through fuel cells, producing only water and heat as by products. Such vehicles are considered to be a zero-emission alternative to conventional internal combustion engine vehicles.

Given the ambitious EU goals for the transport sector, from 2035 onward, only zero-emission cars and light commercial vehicles will be allowed for the registration in the EU. To move toward this goal, there are many supporting policies provided that should accelerate the use of battery electric vehicles and fuel cell vehicles. Due to supporting framework, the number of electric vehicles is rapidly increasing, reaching more than 26 million in 2023. However, the number of fuel cell vehicles

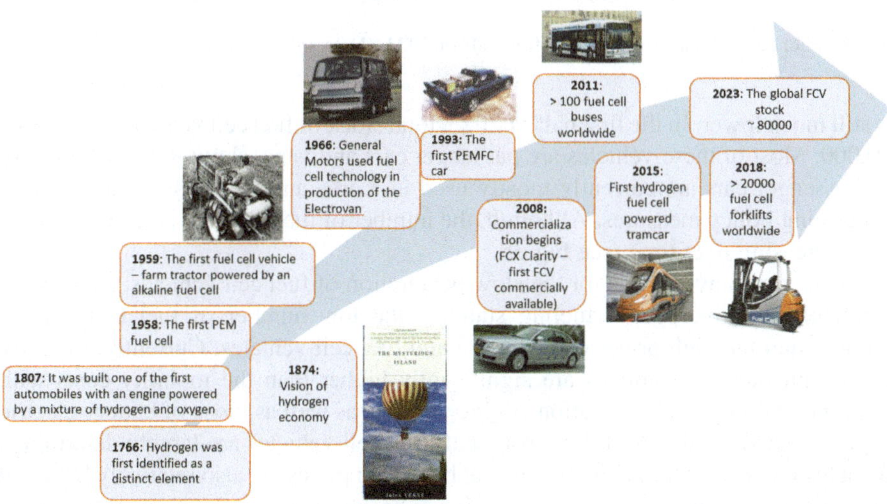

Fig. 6 Major milestones in the hydrogen use in the transport sector. (Adapted from [49])

Fig. 7 Basic elements of fuel cell vehicle

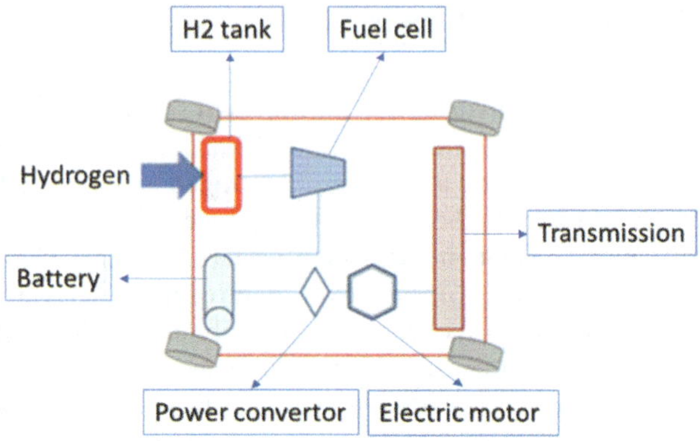

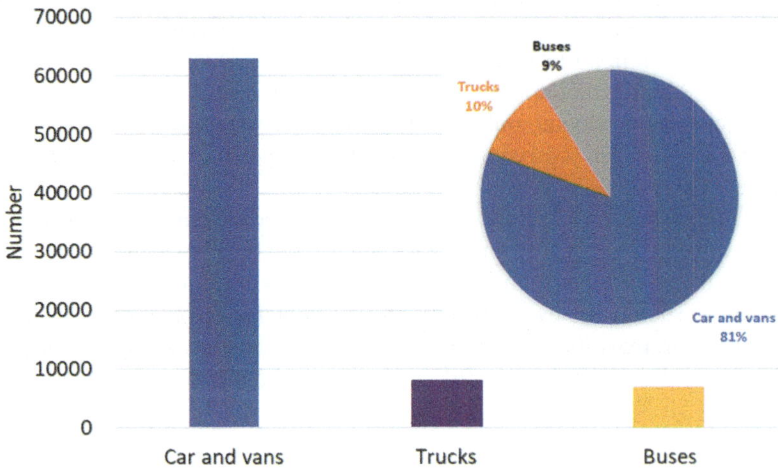

Fig. 8 Fuel cell vehicles by segment, first half of 2023. (Data source [20])

is still much lower. In the first half of 2023, total stock of fuel cell vehicles was below 80,000. Most of these vehicles are passenger cars and vans, about 81% (see Fig. 8).

Passenger cars are currently mostly used in Korea and the United States due to supporting policy measures. Although, the number of fuel cell cars is relatively low, it is increasing over time (see Fig. 9).

There are many reasons for the slow penetration of fuel cell vehicles, such as the availability of hydrogen refueling stations, the low number of fuel cell vehicle models, and the high prices of hydrogen and fuel cell vehicles. Currently, mobility costs with fuel cell vehicles are significantly higher than the mobility costs with conventional internal combustion engine vehicles as well as those of battery electric vehicles (see Fig. 10). Mobility costs with fuel cell vehicles are largely determined by investment costs for vehicles. Current hydrogen prices are also relatively high and different from country to country. Yet, it can be assumed that the price of hydrogen will fall significantly in the future as the offering expands. However, in spite of

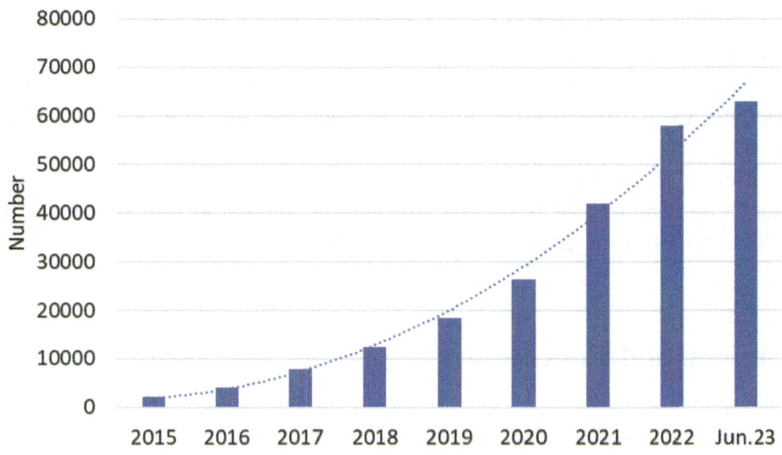

Fig. 9 Development of the global stock of fuel cell cars [20, 24, 32, 50–53]

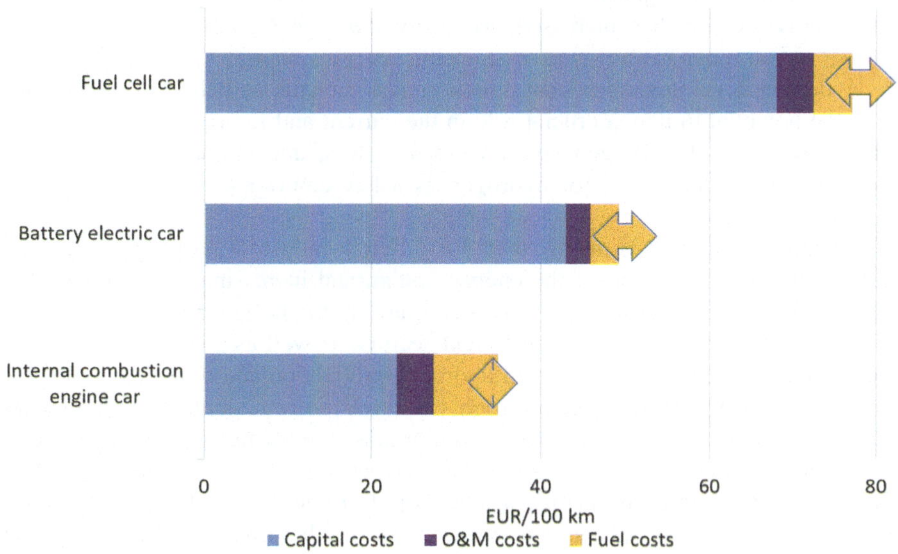

Fig. 10 Mobility cost with different types of cars. (Adapted from [24])

falling prices and increasing number of the fuel cell cars on the market, they can hardly compete with battery electric cars in the common daily use pattern.

One of the advantages of hydrogen is that it can be used in all transport modes. Beside the use in light duty vehicles, there are already good experiences in hydrogen use in buses, trucks, trains, airplanes, forklifts, ferries, etc. Currently, China is mostly focusing on fuel cell heavy-duty vehicles and accounts for more than 95% of total fuel cell trucks and about 85% of global fuel cell buses. In Germany, 36 fuel cell trains have been in operation since last year. In Norway, the use of liquid hydrogen ferry has also started in 2023 [20].

Over the last decades, many countries and companies have invested in the development of hydrogen-fueled aircraft. Empirical data show positive results for various projects.

Yusaf et al. (2024) [54] have provided an overview of the current developments and future applications of hydrogen in aviation, indicating the major barriers to widespread adoption of hydrogen technology in this area.

5 Outlook and Conclusions

The utilization of hydrogen is evolving, with continuous research and development. The focus is put on making hydrogen more cost-effective, more environmentally friendly, and widely applicable in diverse sectors, especially in the industry, transport sector, and power generation. The interest in hydrogen, as a clean energy carrier and energy storage, has increased driven by the global goals to curtail carbon emissions and move toward more sustainable energy system.

Hydrogen is gaining significant attention as a versatile and clean energy carrier with the potential to play a crucial role in the current and future energy landscape. The current use of hydrogen spans various sectors, and ongoing advancements suggest a promising future for hydrogen as a key contributor to sustainable and low-carbon energy systems.

Currently, hydrogen is mostly used for different industrial purposes. However, through the transformation of the energy system and increasing use of renewable energy sources, new prospects for hydrogen are on the horizon. Hydrogen can be used as a clean energy carrier in different sectors, as well as a storage for surplus electricity. Ongoing improvement in hydrogen production and usage technologies, in combination with a growing awareness of environmental challenges, is making hydrogen one the key players in the transition to a more sustainable energy system. However, the future of hydrogen is very dependent on the supporting policy framework and long-term visions, which should provide more clarity on the mid- and long-term developments and reduce risk in the investment in hydrogen technologies.

References

1. Ritchie H, Roser M, Rosado P (2024) Renewable energy. Our World in Data. https://ourworldindata.org/renewable-energy. Accessed 26 Jan 2024
2. Ball M, Weeda M (2015) The hydrogen economy—vision or reality? Int J Hydrog Energy 40: 7903–7919. https://doi.org/10.1016/j.ijhydene.2015.04.032
3. Groll M (2023) Can climate change be avoided? Vision of a hydrogen-electricity energy economy. Energy 264:126029. https://doi.org/10.1016/j.energy.2022.126029

4. Rifkin J (2003) Hydrogen Economy. Penguin Group US, New York
5. Rosen MA, Koohi-Fayegh S (2016) The prospects for hydrogen as an energy carrier: an overview of hydrogen energy and hydrogen energy systems. Energy Ecol Environ 1:10–29. https://doi.org/10.1007/s40974-016-0005-z
6. Capurso T, Stefanizzi M, Torresi M, Camporeale SM (2022) Perspective of the role of hydrogen in the 21st century energy transition. Energy Convers Manag 251:114898. https://doi.org/10.1016/j.enconman.2021.114898
7. Abdin Z, Zafaranloo A, Rafiee A et al (2020) Hydrogen as an energy vector. Renew Sust Energ Rev 120:109620. https://doi.org/10.1016/j.rser.2019.109620
8. Yap J, McLellan B (2023) A historical analysis of hydrogen economy research, development, and expectations, 1972 to 2020. Environments 10:11. https://doi.org/10.3390/environments10010011
9. Majumdar A, Deutch JM, Prasher RS, Griffin TP (2021) A framework for a hydrogen economy. Joule 5:1905–1908. https://doi.org/10.1016/j.joule.2021.07.007
10. Mac Dowell N, Sunny N, Brandon N et al (2021) The hydrogen economy: a pragmatic path forward. Joule 5:2524–2529. https://doi.org/10.1016/j.joule.2021.09.014
11. Akhtar MS, Khan H, Liu JJ, Na J (2023) Green hydrogen and sustainable development—a social LCA perspective highlighting social hotspots and geopolitical implications of the future hydrogen economy. J Clean Prod 395:136438. https://doi.org/10.1016/j.jclepro.2023.136438
12. Dillman KJ, Heinonen J (2022) A 'just' hydrogen economy: a normative energy justice assessment of the hydrogen economy. Renew Sust Energ Rev 167:112648. https://doi.org/10.1016/j.rser.2022.112648
13. Kar SK, Sinha ASK, Harichandan S et al (2023) Hydrogen economy in India: a status review. WIREs Energy Environ 12:e459. https://doi.org/10.1002/wene.459
14. Kar SK, Sinha ASK, Bansal R et al (2023) Overview of hydrogen economy in Australia. WIREs Energy Environ 12:e457. https://doi.org/10.1002/wene.457
15. Trattner A, Klell M, Radner F (2022) Sustainable hydrogen society—vision, findings and development of a hydrogen economy using the example of Austria. Int J Hydrog Energy 47: 2059–2079. https://doi.org/10.1016/j.ijhydene.2021.10.166
16. Schlund D, Schulte S, Sprenger T (2022) The who's who of a hydrogen market ramp-up: a stakeholder analysis for Germany. Renew Sust Energ Rev 154:111810. https://doi.org/10.1016/j.rser.2021.111810
17. Ji M, Wang J (2021) Review and comparison of various hydrogen production methods based on costs and life cycle impact assessment indicators. Int J Hydrog Energy 46:38612–38635. https://doi.org/10.1016/j.ijhydene.2021.09.142
18. Ajanovic A, Sayer M, Haas R (2024) On the future relevance of green hydrogen in Europe. Appl Energy 358:122586. https://doi.org/10.1016/j.apenergy.2023.122586
19. Boretti A, Nayfeh J, Al-Maaitah A (2021) Hydrogen production by solar thermochemical water-splitting cycle via a beam down concentrator. Front Energy Res 9:666191. https://doi.org/10.3389/fenrg.2021.666191
20. IEA (2023) Global hydrogen review. IEA, Paris
21. Droege T (2021) What are the colors of hydrogen? Williams Companies. https://www.williams.com/2021/04/23/what-are-the-colors-of-hydrogen/
22. Bridges T, Merzian R (2019) Hydrogen and climate: Trojan horse or golden goose?: request for input—national hydrogen strategy. The Australia Institute, Canberra
23. Dodgshun J (2020) Hydrogen: clearing up the colours. https://www.enapter.com/newsroom/hydrogen-clearing-up-the-colours
24. Ajanovic A, Sayer M, Haas R (2022) The economics and the environmental benignity of different colors of hydrogen. Int J Hydrog Energy. https://doi.org/10.1016/j.ijhydene.2022.02.094
25. Castigliego J, Stasio T (2021) The "colors" of hydrogen. Applied Economics Clinic. https://aeclinic.org/aec-blog/2021/6/24/the-colors-of-hydrogen. Accessed 26 Jan 2024

26. Strategy& (2020) The dawn of green hydrogen-maintaining the GCC's edge in a decarbonized world. https://www.strategyand.pwc.com/m1/en/reports/2020/the-dawn-of-green-hydrogen/the-dawn-of-green-hydrogen.pdf
27. Dawood F, Anda M, Shafiullah GM (2020) Hydrogen production for energy: an overview. Int J Hydrog Energy 45:3847–3869. https://doi.org/10.1016/j.ijhydene.2019.12.059
28. Zhao G, Kraglund MR, Frandsen HL et al (2020) Life cycle assessment of H2O electrolysis technologies. Int J Hydrog Energy 45:23765–23781. https://doi.org/10.1016/j.ijhydene.2020.05.282
29. IEA (2019) The future of hydrogen: seizing today's opportunities. IEA, Paris
30. Hu K, Fang J, Ai X et al (2022) Comparative study of alkaline water electrolysis, proton exchange membrane water electrolysis and solid oxide electrolysis through multiphysics modeling. Appl Energy 312:118788. https://doi.org/10.1016/j.apenergy.2022.118788
31. Longden T, Beck FJ, Jotzo F et al (2022) 'Clean' hydrogen?—comparing the emissions and costs of fossil fuel versus renewable electricity based hydrogen. Appl Energy 306:118145. https://doi.org/10.1016/j.apenergy.2021.118145
32. Noussan M, Raimondi PP, Scita R, Hafner M (2021) The role of green and blue hydrogen in the energy transition—a technological and geopolitical perspective. Sustainability 13:298. https://doi.org/10.3390/su13010298
33. IEA (2018) Hydrogen production costs using natural gas in selected regions, statistics. IEA. https://www.iea.org/data-and-statistics/charts/hydrogen-production-costs-using-natural-gas-in-selected-regions-2018-2
34. BNEF (2020) Hydrogen economy outlook: key messages. BNEF, New York
35. Hydrogen Council (2020) Path to hydrogen competitiveness: A cost perspective. Hydrogen Council, Brussels
36. European Commission (2023) Hydrogen. https://energy.ec.europa.eu/topics/energy-systems-integration/hydrogen_en. Accessed 18 July 2023
37. Dolci F (2018) Green hydrogen opportunities in selected industrial processes – workshop summary report. EUR 29637 EN. Publications Office of the European Union, Luxembourg. https://doi.org/10.2760/634063
38. Madeddu S, Ueckerdt F, Pehl M et al (2020) The CO_2 reduction potential for the European industry via direct electrification of heat supply (power-to-heat). Environ Res Lett 15:124004. https://doi.org/10.1088/1748-9326/abbd02
39. Egeland-Eriksen T, Hajizadeh A, Sartori S (2021) Hydrogen-based systems for integration of renewable energy in power systems: achievements and perspectives. Int J Hydrog Energy 46: 31963–31983. https://doi.org/10.1016/j.ijhydene.2021.06.218
40. IRENA (2022) Geopolitics of the energy transformation: the hydrogen factor. International Renewable Energy Agency, Abu Dhabi
41. Rosenow J (2022) Is heating homes with hydrogen all but a pipe dream? An evidence review. Joule 6:2225–2228. https://doi.org/10.1016/j.joule.2022.08.015
42. Liebreich M (2022) Liebreich: the unbearable lightness of hydrogen. BloombergNEF. https://about.bnef.com/blog/liebreich-the-unbearable-lightness-of-hydrogen/. Accessed 7 May 2023
43. Shell Global (2023) The energy security scenarios. Shell Global. https://www.shell.com/energy-and-innovation/the-energy-future/scenarios/the-energy-security-scenarios.html. Accessed 6 May 2023
44. Bard J, Gerhardt N, Selzam P, et al (2022) The limitations of hydrogen blending in the european gas grid a study on the use, limitations and cost of hydrogen blending in the European gas grid at the transport and distribution level. Fraunhofer Institute for Energy Economics and Energy System Technology (IEE) Kassel
45. IRENA (2022) Global hydrogen trade to meet the 1.5° C climate goal: part II—technology review of hydrogen carriers. International Renewable Energy Agency, Abu Dhabi
46. Ajanovic A (2013) Renewable fuels—a comparative assessment from economic, energetic and ecological point-of-view up to 2050 in EU-countries. Renew Energy 60:733–738. https://doi.org/10.1016/j.renene.2013.06.012

47. van Renssen S (2020) The hydrogen solution? Nat Clim Chang 10:799–801. https://doi.org/10.1038/s41558-020-0891-0
48. Blue motion history of hydrogen cars. https://web.itu.edu.tr/celikmuhamm/bil103/company/webpage.html. Accessed 26 Jan 2024
49. Ajanovic A, Haas R (2021) Prospects and impediments for hydrogen and fuel cell vehicles in the transport sector. Int J Hydrog Energy 46:10049–10058. https://doi.org/10.1016/j.ijhydene.2020.03.122
50. IEA (2019) Global EV outlook 2019. OECD, Paris
51. IEA (2020) Global EV outlook 2020. IEA, Paris
52. TCP IA (2019) AFC TCP 2019 survey on the number of fuel cell vehicles, hydrogen refueling stations and targets. https://www.ieafuelcell.com/fileadmin/publications/2019-04_AFC_TCP_survey_status_FCEV_2018.pdf
53. IEA (2021) Hydrogen e analysis. IEA. https://www.iea.org/energy-system/low-emission-fuels/hydrogen. Accessed 26 Jan 2024
54. Yusaf T, Faisal Mahamude AS, Kadirgama K et al (2024) Sustainable hydrogen energy in aviation—a narrative review. Int J Hydrog Energy 52:1026–1045. https://doi.org/10.1016/j.ijhydene.2023.02.086

Index

A. Kovač (ed.), *Challenges and Solutions in the Hydrogen Value Chain*,
https://doi.org/10.1007/978-3-031-69051-8

R

T

W

SPRINGER NATURE

GPSR Compliance

The European Union's (EU) General Product Safety Regulation (GPSR) is a set of rules that requires consumer products to be safe and our obligations to ensure this.

If you have any concerns about our products, you can contact us on ProductSafety@springernature.com

In case Publisher is established outside the EU, the EU authorized representative is:

Springer Nature Customer Service Center GmbH
Europaplatz 3
69115 Heidelberg, Germany

The manufacturer's authorised representative in the EU is Springer Nature Customer Service Centre GmbH, Europaplatz 3, 69115 Heidelberg, Germany. If you have any concerns regarding our products, please contact ProductSafety@springernature.com

Printed and bound by CPI Group (UK) Ltd, Croydon, CR0 4YY
28/09/2026
02224081-0017